Recycled Waste Materials in Concrete Construction:

Emerging Research and Opportunities

Jahangir Mirza
Research Institute of Hydro-Québec, Canada

Mohd Warid Hussin
Universiti Teknologi Malaysia, Malaysia

Mohamed A. Ismail
Miami College of Henan University, China

A volume in the Advances in Civil and Industrial Engineering (ACIE) Book Series

IGI Global
DISSEMINATOR OF KNOWLEDGE

Published in the United States of America by
IGI Global
Engineering Science Reference (an imprint of IGI Global)
701 E. Chocolate Avenue
Hershey PA, USA 17033
Tel: 717-533-8845
Fax: 717-533-8661
E-mail: cust@igi-global.com
Web site: http://www.igi-global.com

Library of Congress Cataloging-in-Publication Data

Names: Mirza, Jahangir, 1944- author. | Hussin, Mohd Warid, 1951- author. | Ismail, Mohamed A., 1968- author.
Title: Recycled waste materials in concrete construction : emerging research and opportunities / by Jahangir Mirza, Mohd Warid Hussin, and Mohamed A. Ismail.
Description: Hershey, PA : Engineering Science Reference, [2019] | Includes bibliographic references.
Identifiers: LCCN 2018050685| ISBN 9781522583257 (hardcover) | ISBN 9781522583264 (ebk.)
Subjects: LCSH: Waste products as building materials. | Concrete construction--Environmental aspects.
Classification: LCC TA403.6 .M576 2019 | DDC 624.1/8340284--dc23 LC record available at https://lccn.loc.gov/2018050685

This book is published in the IGI Global book series Advances in Civil and Industrial Engineering (ACIE) (ISSN: 2326-6139; eISSN: 2326-6155)

British Cataloguing in Publication Data
A Cataloguing in Publication record for this book is available from the British Library.

All work contributed to this book is new, previously-unpublished material.
The views expressed in this book are those of the authors, but not necessarily of the publisher.

For electronic access to this publication, please contact: eresources@igi-global.com.

Advances in Civil and Industrial Engineering (ACIE) Book Series

ISSN:2326-6139
EISSN:2326-6155

MISSION

Private and public sector infrastructures begin to age, or require change in the face of developing technologies, the fields of civil and industrial engineering have become increasingly important as a method to mitigate and manage these changes. As governments and the public at large begin to grapple with climate change and growing populations, civil engineering has become more interdisciplinary and the need for publications that discuss the rapid changes and advancements in the field have become more in-demand. Additionally, private corporations and companies are facing similar changes and challenges, with the pressure for new and innovative methods being placed on those involved in industrial engineering.

The **Advances in Civil and Industrial Engineering (ACIE) Book Series** aims to present research and methodology that will provide solutions and discussions to meet such needs. The latest methodologies, applications, tools, and analysis will be published through the books included in **ACIE** in order to keep the available research in civil and industrial engineering as current and timely as possible.

COVERAGE

- Transportation Engineering
- Urban Engineering
- Structural Engineering
- Hydraulic Engineering
- Production Planning and Control
- Materials Management
- Operations research
- Ergonomics
- Quality Engineering
- Coastal Engineering

The Advances in Civil and Industrial Engineering (ACIE) Book Series (ISSN 2326-6139) is published by IGI Global, 701 E. Chocolate Avenue, Hershey, PA 17033-1240, USA, www.igi-global.com. This series is composed of titles available for purchase individually; each title is edited to be contextually exclusive from any other title within the series. For pricing and ordering information please visit http://www.igi-global.com/book-series/advances-civil-industrial-engineering/73673. Postmaster: Send all address changes to above address.

Titles in this Series

For a list of additional titles in this series, please visit:
https://www.igi-global.com/book-series/advances-civil-industrial-engineering/73673

Optimizing Current Strategies and Applications in Industrial Engineering
Prasanta Sahoo (Jadavpur University, India)
Engineering Science Reference • ©2019 • 382pp • H/C (ISBN: 9781522582236) • US $245.00

Big Data Analytics in Traffic and Transportation Engineering Emerging Research...
Sara Moridpour (RMIT University, Australia)
Engineering Science Reference • ©2019 • 188pp • H/C (ISBN: 9781522579434) • US $165.00

Optimization of Design for Better Structural Capacity
Mourad Belgasmia (Setif 1 University, Algeria)
Engineering Science Reference • ©2019 • 283pp • H/C (ISBN: 9781522570592) • US $215.00

Reusable and Sustainable Building Materials in Modern Architecture
Gülşah Koç (Yildiz Technical University, Turkey) and Bryan Christiansen (Global Research Society, LLC, USA)
Engineering Science Reference • ©2019 • 302pp • H/C (ISBN: 9781522569954) • US $195.00

Measuring Maturity in Complex Engineering Projects
João Carlos Araújo da Silva Neto (Magnesita SA, Brazil) Ítalo Coutinho (Saletto Engenharia de Serviços, Brazil) Gustavo Teixeira (PM BASIS, Brazil) and Alexandro Avila de Moura (Paranapanema SA, Brazil)
Engineering Science Reference • ©2019 • 277pp • H/C (ISBN: 9781522558644) • US $245.00

Big Data Analytics for Smart and Connected Cities
Nilanjan Dey (Techno India College of Technology, India) and Sharvari Tamane (Jawaharlal Nehru Engineering College, India)
Engineering Science Reference • ©2019 • 348pp • H/C (ISBN: 9781522562078) • US $225.00

Contemporary Strategies and Approaches in 3-D Information Modeling
Bimal Kumar (Glasgow Caledonian University, UK)
Engineering Science Reference • ©2018 • 313pp • H/C (ISBN: 9781522556251) • US $205.00

For an entire list of titles in this series, please visit:
https://www.igi-global.com/book-series/advances-civil-industrial-engineering/73673

IGI Global
DISSEMINATOR OF KNOWLEDGE

701 East Chocolate Avenue, Hershey, PA 17033, USA
Tel: 717-533-8845 x100 • Fax: 717-533-8661
E-Mail: cust@igi-global.com • www.igi-global.com

Table of Contents

Preface

Recycled Waste Materials in Concrete Construction: Emerging Research and Opportunities is the first edition of the book comprising of 10 chapters. Each of these chapters discusses a different type of industrial, agricultural and/or natural wastes; some have been already utilized in the concrete industry while many others demonstrate promise for future use. Recycling of those waste materials, found in abundance, not only wards off deleterious environmental hazards, but also have been known to actually produce wealth by adding value through ecology.

Chapter 1 deals with the characterization and assessment of iron ore tailings as raw materials for construction industries. It specifies the production process of Iron Ore and generation of waste material followed by listing the nature and prospective issues of Iron Ore Tailings (IOT). Methods of IOT characterization are explained through the five elements which are: chemical composition, leaching behavior, thermal stability, mineralogical characterization and morphology. The experimental program and research results of this study are explained in 6 subtitles namely; chemical composition; leaching behaviour; thermal stability; X-Ray diffraction pattern; Fourier transform infrared spectroscopy (FTIR); and field emission scanning electron microscopy (FESEM/EDX). Conclusions are given to highlight the findings of the research.

Effect of high volume palm oil biomass waste used in mortar is discussed in the second chapter. The reader is given an introduction of palm oil biomass waste, as well as palm oil industry issues regarding its by-products waste are explored. It is followed by a clear description of each waste and its effect when added to mortar mixes. Furthermore, a research study on the effect of palm oil fuel ash, palm oil kernel shell, palm oil fibre on mortar properties was carried out and the experimental program details are given under 4 subtitles. Splitting tensile strength and flexural strength were performed to test the engineering properties of mortar containing different types of waste. Results and discussion are provided for additional grasp. A summary of research outcomes is then given to highlight the potential of these wastes to be used in industry.

Chapter 3 is based on performance of mortar grouts incorporating rice husk ash (RHA) and fly ash (FA). Detailed experimental work was conducted to investigate the mechanical properties of mortar grouts using RHA and FA as partial replacement of cement. It analysed the compressive strength and durability of mortar grouts in its hardened state. Durability tests such as water absorption, apparent volume of permeable voids, sorptivity and rapid chloride penetration tests are researched. Detailed results and discussion which focused on mechanical properties as well as durability of hardened state mortar grouts are presented, followed by a conclusive summary of the study's outcome.

Chapter 4 deals with the effects of metakaolin (MK) on geopolymer industry. The research topics of MK-based geopolymer cover reaction mechanisms and kinetics. This chapter aims at augmenting knowledge about enhancing mechanical properties of geopolymer mortars/concrete using MK. Specifically; this chapter presents literature studies as well as current experimental studies which delineate the effect of MK on fresh- and hardened-state properties of geopolymer mortars (GPMs). Properties and characteristics of metakaolin are explained followed by properties of fresh MK Mortars. Properties of hardened MK concrete and durability properties of MK mortars are explained. Applications of MK based geopolymers and metakaolin based geopolymers as repair materials are also included in this chapter. Brief conclusions drawn from the study are listed toward the end.

A conspectus based upon a compelling topic, namely, versatility of cockle shell use in concrete, which is yet to be investigated, is covered in Chapter 5. An introduction to enlighten the reader with this promising waste material precedes a review of environmental issues with cockle shell. Cockle Trade is an important subtitle that covers Cockle Shell waste generation, research and development related to the deployment on the use of Cockle Shell, Processing Cockle Shell for making construction material is given in details. It covers its use as partial coarse aggregates replacement and partial sand replacement. Outcomes of the study are summarized towards the end.

Chapter 6 deals with the durability performance of mortar containing ceramic tile waste exposed to sulphate attack. The introduction discusses the latest development regarding this subject as no case study has been found where ceramic tile waste was actually used in the field. Sulphate resistance, visual appearance, mass change and residual compressive strength are studied as an experimental program. Microstructure analysis to verify the results and cement the conclusions was explained in details. Finally a comprehensive summary of the research output is also included.

The effects of palm oil fuel ash as micro-filler on interfacial porosity of polymer concrete is the subject of Chapter 7. Beginning with a brief introduction about the topic, the materials and method used in this study are explained. Results and discussion about the research program are given explicitly and conclusions are drawn.

Chapter 8 evaluates a most interesting and up-to-date topic of electric arc furnace oxidizing slag aggregates' quality and development of functional concrete. A comprehensive introduction to the subject is given followed by a scientific method of stabilizing electric arc furnace (EAF). Oxidizing slag (EOS) is explained in brief. Subsequently, expansion mechanism of EOS and physical and chemical properties of EOS aggregates are covered in reasonable detail. A method for quantitative evaluation of free CaO contained in EOS that covers free CaO content as a function of aging period and open storage position for the EOS and ERS samples are explained. Properties of functional concrete using EOS aggregates are described in detail through the experimental program and test results. Finally brief conclusions are provided for the readership.

Chapter 9 covers a comparison between ordinary Portland cement and geopolymer concretes against sulphuric acid attack. An intensive introduction to the topic is given. Methodology of the experimental program, with emphasis on preparation of materials and mix design is described. Testing procedure of GSCC is given and durability test for sulphuric acid and cost analysis are briefly explained. Conclusions of this chapter would help give a quick analysis of this chapter's details.

Self-healing mortar is the title of the last chapter of this book. A detailed introduction about self-healing mortar gives the readers an adequate background. A problem statement that summaries the issue is given afterwards. Thermoset Polymer – Diglycidyl Ether of Bisphenol A Type of Epoxy Resin is explained briefly. Self-healing concept is explained in details then followed by the experimental program results of compressive strength, damage degree of concrete after self-healing, permeability test, reaction between Epoxy Resin and Hydroxyl Ion. A brief summary of the findings is given at the end.

The versatility of this book, compared to others, lies in its timely compilation about the most significant development of the 20th century (i.e., concrete from waste materials). The editors believe no such book currently exists which compiles information about an extensive variety of wastes that could be used in the concrete industry to generate low cost environmental friendly materials. Moreover, a few chapters reveal a combination of these wastes, new approaches to old materials and unique demands related to waste materials. The availability of the book to engineers, technologists, researchers, contractors, consulting firms, government agencies dealing with construction, transportation and environment, general public, etc., is very crucial.

The results presented in these studies are not only beneficial both economically and environmentally, they also provide the concrete industry with technical information about valuable resources which play a key role in meeting the challenges of sustainable construction in today's world. The high demand of natural resources due to rapid urbanization and the disposal problem of industrial and agricultural wastes in developed and in developing countries have created opportunities for utilizing these wastes in concrete. Many of them have already been used in concrete as additive or replacement to cement, fine aggregates and coarse aggregates. By doing so, these wastes drastically improve many properties of fresh and hardened concrete, paving the way for major developments in concrete and construction industries.

The principal binder in concrete is Portland cement whose production requires exorbitant amount of energy consumption, is costly and a major contributor to green-house gases (GHG; one ton of Portland cement releases approximately one ton of CO_2). Furthermore, it consumes huge quantities of virgin materials that cause depletion of natural resources, such as forests, hill, mountains, etc. at an alarming rate. Given that these challenges must be dealt with effectively, the commitment to deploy immense quantities of industrial, agricultural and natural waste materials (palm-oil fuel ash, fly ash, coal and oil-burning by-products, bottom ash, rice husk ash, bagasse ash, metakaolin, used tires, cement dust, stone crushers dust, marble dust, silica fume, glass, etc.) becomes imperative. Research has adequately shown that waste materials incorporated in concrete could enable the construction industry to not only become much more sustainable and economical, but also help overcome other related ecological, sociological and economic problems associated with it.

Effective utilization of various waste materials in the concrete and construction industry whose growth knows no boundaries and mounting evidence of worldwide interest suffice the need to produce a collective anthology of a wide variety of waste materials available today.

Jahangir Mirza
Research Institute of Hydro-Quebec, Canada

Mohd Warid Hussin
Universiti Teknologi Malaysia, Malaysia

Mohamed A. Ismail
Miami College of Henan University, China
January 20, 2019

Chapter 1
Characterization and Assessment of Iron Ore Tailings as Raw Materials for Construction Industries

ABSTRACT

This chapter deals with the characterization and assessment of iron ore tailings (IOT) as raw materials for the construction industry. This chapter specifies the production process of iron ore and generation of waste material followed by listing the nature and prospective issues of IOT. Methods of IOT characterization are explained through five elements, which are chemical composition, leaching behavior, thermal stability, mineralogical characterization, and morphology. The experimental program and research results of this study are explained in six subtitles, namely chemical composition, leaching behaviour, thermal stability, x-ray diffraction pattern, Fourier transform infrared spectroscopy (FTIR), and field emission scanning electron microscopy (FESEM/EDX). Results revealed that the IOT materials are suitable for use in construction and building industries due to their substantial silica and alumina contents and could possibly be used to fabricate paving blocks, sand-crete blocks, mud blocks, geopolymer bricks, and ceramic floor tiles.

DOI: 10.4018/978-1-5225-8325-7.ch001

INTRODUCTION

Iron is one of the world's most commonly used metal derived from iron ore. Iron is the key ingredient, representing almost 95% of all metals used per year (U.S. Geological Survey [USGS], 2017). It is used primarily in structural engineering applications, maritime structures, automobiles, general industrial machineries and equipment, etc.

World production of iron ore was more than two billion metric tonnes of raw ore annually. China is by far the largest producer, consumer, and importer of iron ore. In 2015 alone China produced 1.3 billion tonnes of iron ore equivalent to 44% of the world's output. Australia comes as the second largest producer of iron ore that produced 824 million tonnes in 2015 representing more than 20% of the global output. Brazil is the third largest producer of iron ore producing 751 and 411 million tonnes in 2015 and 2014, respectively. Brazil's 2015 output represented 12% of the world's production. In the past, India was a world leader, but now stands the fourth largest producer. In 2015, India produced 129 million tonnes of iron ore. Russia is the fifth largest producer, and in 2015 it produced 112 million tonnes of crude iron ore.

Iron ore tailings (IOT) are waste materials generated when iron ore is processed by separating valuable fractions from the worthless ore. Enormous quantities of industrial wastes produced from mining industries are usually disposed of to landfills due to uneconomic attractive usage. The continuous disposal of IOT endures over-burden to the mining industries and the community in terms of environmental and economic perspectives. For instance, in China the total stockpiles of iron ore tailings was 2.899 billion metric tonnes during the years between 2007 and 2011. Of that, approximately 806 million metric tonnes were generated in 2011. However, only 307 million metric tonnes of tailings were comprehensively recycled (Cai et al., 2016). Goyal et al. (2015) reported that approximately 10 – 12 million tonnes of such mined ores are lost as tailings in India. In Europe, about 2.75 billion tonnes of mining waste have been produced between 1998 and 2001 (Grangeia et al., 2011). The estimated iron ore production in Brazil for 2015 was 751 million tonnes and generated 260 million tons of iron-ore tailings (Carrasco et al., 2017). The increased accumulation of iron ore tailings and eventually their disposal pose an over-burden to the mining industries and the community in terms of environmental and economic perspective (Zhang et al., 2006). Lack of space for the disposal of huge amount of IOT stocked in the industry will become a major problem in the future. There are other possible issues, such as, leaching of heavy metals and acid mine drainage which could cause havoc to the community and the environment (Hitch et al., 2010).

As mentioned earlier, IOT stockpiled in the tailing dams might be risky to the environment and the impacts could be physical, chemical or geotechnical instability. The possible effects for storing the tailings in the dam could lead to soil and surface water pollution due to toxic substances such as lead, sulphates and dissolved metals. Sulphates, in particular are susceptible to chemical oxidation when exposed to oxygen and form acid in the soil. In this case the acid might possibly kill vegetation and diffuse into ground water.

The persistent disposal of IOT in landfills or tailing dams has a range of environmental issues, which include: erosion, dust, water and soil pollution, negative effects on the ecosystems and loss of land fertility. The difficult situations that might be encountered are during the failure of tailing dams or collapse of heaps due to earthquakes and heavy rainfall, which could affect the environment and health safety of human life (Cai et al., 2011).

In order to find a solution for the environmental issues raised and sustainability of natural resources, there is certainly a need to further study the characterization and microstructure of the waste material that will increase the percentage of tailings utilization as construction and building materials. Current utilization of IOT at 7 to 10% (Zhao et al., 2014; Huang et al., 2013b) is very low compared to huge disposal ranging from 5 to 7 billion tonnes per year worldwide (Edraki et al., 2014). Increasing the utilization of IOT will provide eco-friendly, economic and environmentally sustainable mining industries and also provide alternative and cheaper building and construction materials.

For this study, the IOT was obtained from Kota Tinggi iron ore mills in Johor; a southern state in Malaysia Peninsula. Physical test results showed that it had a specific gravity of 2.6, bulk density of 1710 kg/m^3, fineness modulus of 2.07 and water absorption rate of 7.0%.

Production of Iron Ore and Generation of Waste Material

The process of mining starts with assembling the ores from excavation, blasting and exploration processes. These processes eventually produced wastes that were commonly found as mine overburden wastes, rocks and top soil of different scales. At the end of these processes, mine ore is produced which might be of different types depending upon the area of ore deposit. The ore will further undergo other processes through pulverizing, upgrading, leaching, and concentration (Wills and Napier-Munn, 2005) for efficient and maximum liberation. At the time of pulverization, solid ore materials are crushed, grinded and cut down into smaller scale. This process will strip the valuable particles of the minerals and raise the appearance in order to continue the next process. The valuables minerals are packed and shut into different arrangement or blended with other materials. The pulverization method of processing assists to release the valuable and worthwhile ore minerals from the different arrangements. The pulverization method of processing will also produce spoils or fines as waste materials which are considered worthless.

In order to achieve higher percentage of metals or precious mineral, the mineral ores are required to undergo another process using upgrading and concentration of the ore through separation of the precious minerals from the worthless type. The upgrading and concentration are performed through magnetic separation, chemical separation, washing and froth flotation relying on the mineral ore being treated. The processing of iron ore ends at this phase but gold and other ores will undergo leaching processes which were performed through subjecting the ore to upgraded and concentrated cyanidation (Committee on Technologies for the Mining Industry, 2002).

The stages of ore processing as enumerated produce huge wastes which include; leached residues, slurries, processed waste water, tailings and hazardous chemicals. The product of the ore processing at this stage is the formation of bulk raw material. However, the bulk raw material will further undergo refining and smelting before finally being accepted as marketable product. Waste materials such as ashes and slags are also produced during the processes of refining and smelting.

Nature and Prospective Issues of IOT

Persistent disposal of iron ore tailings unleashed problems for the community as well as mining companies in terms of environmental and economic issues. The environmental issue of pollution is the main concern to the mining industries. The acquisition of nearby landfill space for the disposal of tailings through friendly and sustainable means is another main concern of the mining industries. When the tailings are successfully disposed off, it is necessary to keep on monitoring and managing the disposal site (McKinnon, 2002). That will ensure the prevention of any erosion possibility in case of dam failure while acid mine drainage will be curtailed (Grangeia et al., 2011;Yellishetty et al., 2008). The main prospective issue of the tailings is that they might release heavy metals and hazardous toxic elements into the environment (Ye et al., 2014). Similarly, the mining industries were also faced with the problems of cost for wastes mine disposal which cause decline in huge amount of production profit.

In the past, mining industries were engaged in discharging their tailings directly into the sea (McKinnon 2002). Presently, some of these tailings are pressurized into tailing dams as slurry while others are stored as wastes near the mine sites (Wills and Napier-Munn, 2005). Some of the mining industries utilize certain part of the tailings as backfill and land reclamation to refill previous area in open cast mines and quarries that were dug-out (Skarżyńska, 1995).

In order to handle the problems of mine tailings that satisfy community and devise sustainable mining techniques, it becomes mandatory to assess the various methods of managing such tailings produced from the industries. In this case, the economic and sustainable options of handling mine tailings are obligatory. The best reuse alternative for these tailings is to use it in concrete production where large volumes could be utilized. Literature studies have shown that some of these tailings could be used in the construction of buildings and civil engineering infrastructures (Skarżyńska, 1995;Choi et al., 2009).

Methods of Iron Ore Tailings Characterization

The characteristics of IOT can vary greatly and depend on the ore mineralogy together with the physical and chemical processes used to extract the economic product. The tailings characteristics must be determined to establish their long-term behavior and the potential short and long-term liabilities as well as environmental impacts. Once the likely characteristics of the tailings are determined from mineralogical examinations, the necessary design requirements can be identified to mitigate and render possible use in paving block fabrication, sand-crete block, mud block, geopolymer brick or ceramic floor tiles as construction and building materials. To help determine the characterization and assessments of tailings, the following methods were utilized:

1. Chemical composition
2. Leaching behavior
3. Thermal stability

4. Mineralogical characterization
5. Morphology

The engineering characteristics of tailings are, in most instances, influenced by the method of deposition and materials' characteristics.

Chemical Composition

The IOT were tested for chemical oxide composition using S4 Pioneer X-ray fluorescence spectroscopy (XRF) machine. The XRF machine works by irradiating the samples with an intense X-ray beam from a radioisotope source. The primary source of the rays excites the sample by detaching the tightly bound electrons' inner shell from their orbits in the samples of excited atoms. When the excited atoms are relaxed to their original state, fluorescent X-rays are emitted. The energy from the emitted rays detected is used for the identification of elements in the sample using an energy dispersive detector; while the intensity of the X- rays is used to determine the quantity of the elements.

Leaching Behaviour

Toxicity Characteristic Leaching Procedure (TCLP) was used in determining the concentration of heavy metals in IOT. The test was performed according to the US EPA Method 1311 (1992) whereby IOT materials less than 2.00 mm in size were mixed with deionized water at a liquid - solid ratio of 20:1. Two test samples were agitated with a speed of 30 rpm for 24h. After the extraction and filtration of the leachates, heavy metal ion concentrations therein were determined by means of an Inductively Coupled Plasma–Mass Spectrometry (ICP–MS).

Thermal Stability

This aspect of the test evaluates the thermal stability of the materials due to thermal decomposition. IOT were subjected to thermo-gravimetric and differential analysis (TGA/DTA) test. Test specimens were prepared and tested in powder forms prepared by grinding. The equipment used for the test is Mettler Toledo TGA/DSC thermo-gravimetric analyser. The specimens were heated within a temperature range of 30.0 ± 2.0 °C to 1000 ± 2.0 °C at a heating rate of 10 °C min^{-1} and helium content of 10ml/min. Within the specified temperature range, the mass loss and the decomposition temperatures were determined, recorded and reported.

Mineralogical Characterization

The Mineralogical characterization of IOT was determined using X-ray diffraction analysis (XRD) and Fourier Transform Infrared Spectroscopy (FTIR). XRD enables the determination of the crystalline structures, the quantitative and qualitative analysis phase, the study of phase transformations, the crystallographic texture, the size of crystallites and the internal stresses in a variety of materials. In this study, a SIEMENS Diffractometer D5000 with X-ray source of Cu Kα radiation was used. The scan step was 0.02° using a scanning rate of 0.5 °C/min and in the range 2-theta-scale from 5 ° to 80 °. The scale on the x-axis (diffraction angle) of the usual XRD pattern gives the crystal lattice spacing and the y-axis scale (peak height) shows the intensity of the diffraction rays. The diffraction samples were powdered by grinding.

FTIR is an instrument used to obtain an infrared spectrum of absorption or transmission of a solid, liquid or gas. The objective of any absorption or transmission of spectroscopy is to measure the wavelength, which determines the material's molecular characterization. FTIR experiment was performed on IOT samples to examine the absorption of infrared spectrum. FTIR was used to identify the chemical functionality and materials' molecular characterization. The spectra were recorded from 400 to 4000 cm^{-1} at a resolution of 4 cm^{-1} with an accumulation of 16 scans. A transparent disk (pallet) was used for the test. The pallet was prepared by mixing about 1 mg of the test specimen with about 100 g of potassium bromide (KBr) in an agate mortar and then pressed using a pallet dye under pressure. Nitrogen (N_2) gas was used to purge the spectrometer.

Morphology

Morphological and microscopic analysis of IOT was determined by using Field Emission Scanning Electron Microscopy (FESEM). The FESEM device was GEMINI FESEM versatile ultrahigh resolution with a variable pressure solution, equipped with an energy dispersive X-ray analyser (EDX) unit. The EDX was used to analyse the elemental composition of the respective materials.

RESULTS AND DISCUSSION

Chemical Composition

The chemical composition of IOT was determined by using X- Ray Fluorescence spectroscopy (XRF) as shown in Table 1. The major component of IOT is silicon dioxide or silica (SiO_2). The mass content of silica was 56% while alumina (Al_2O_3) content was 10% and iron oxide (Fe_2O_3) content was 8.3%. The silica usually reacts with calcium hydroxide to produce more aggregate binding gel - CSH (Gite et al., 2013). Fe_2O_3 is the main colorant in the mine tailing, being responsible for the reddish colour. The calcium oxide (CaO) content in IOT was 4.3%, which could possibly be taking part in the pozzolanic reaction and creating extra C-S-H gels and is beneficial for strength development of the concrete. Interestingly, no trace of sulphur trioxide was found in the chemical composition. Although various researchers have raised suspicion about the possibility of corrosion effect on the IOT material due to the high content of Fe_2O_3 (Kuranchie et al., 2015;Yellishetty et al., 2008), the results of chemical analysis indicated that the IOT is very suitable for construction and building materials.

Leaching Behaviour

The heavy metals leaching test results are listed in Table 2 along with the regulatory limits. As reported in chemical composition, IOT contain certain amount of heavy metals that can be leached by water, but the concentrations of As, Ba, Cd, Cr, Pb and Zn in the leachate were below the allowable regulatory limit for waters (groundwater, surface water and tap water) and farmland soil based on the requirement of U.S EPA (2000) standards. Huang et al. (2013a) observed similar trend when they investigated the leaching behavior of IOT powder as cement replacement. They reported that all the leaching contaminants were below the regulatory limits. Similarly, da Silva et al (2015) also reported that chromium, lead and cadmium elements were below the detection limits of the Inductive Couple Plasma (U.S EPA

Table 1. Chemical composition of IOT

Chemical Composition (%)	SiO_2	Al_2O_3	Fe_2O_3	CaO	MnO	K_2O	TiO_2	CUO	PbO	LOI
IOT	56	10	8.3	4.3	1.7	1.5	0.4	0.2	0.4	3.3

Table 2. Concentration of heavy metals in IOT

Elements	U.S EPA Test Method 1311 (1999) Regulatory Limits (mg/l)	Concentration in IOT (mg/l)
As	0.05	0.0013
Ba	1.00	0.0191
Cd	0.01	0.0011
Cr	0.05	0.0047
Pb	0.05	0.0009
Se	0.01	0.0010
Ag	0.05	0.0094
Zn	0.50	0.0003
Cu	0.002	0.0010

1992), when the leaching concentration of concrete blocks containing IOT was examined. They also reported that other heavy metals that could affect the environment, such as water, soil and groundwater contamination were not detected. Related studies by Coruh and Ergun (2006) and Hashem et al. (2011) have also shown that heavy metals contained in mine tailings can be effectively immobilized in cement mixtures. Therefore, IOT used in this study can be considered as a non-hazardous mine waste material.

Thermal Stability

The thermal stability in terms of thermo-gravimetric and differential thermal analysis (TGA/DTA) curves of IOT is shown in Figure 1. It can be seen that the TGA curves showing peaks ranging between 40 to 230 °C, and 260 to 600 °C are normally associated with loss of surface water and dihydroxylation of the material (Maifala and Tabbiruka, 2007). The mass loss due to dihydroxylation was completed at 570 °C. The differential thermal analysis (DTA) curve shows two sharp exothermic peaks at 280 °C and 580 °C corresponding to the loss of surface water and to dihydroxylation of the IOT material. The small endothermic peaks between 200 °C and 580 °C are due to loss of structural water as discussed by Temuujin et al., (1999). TGA curve further showed that the sample decomposed from 100% to 91.14%, which correspond to the mass loss of 6.15% and 2.71% due to the decomposition of adsorbed water between 40 to 230 °C and 260 to 600 °C respectively, making a total loss of about 8.86%. This could be due to high porosity and loose nature of the particles.

Figure 1. Thermo-gravimetirc and thermal differential analyses of IoT

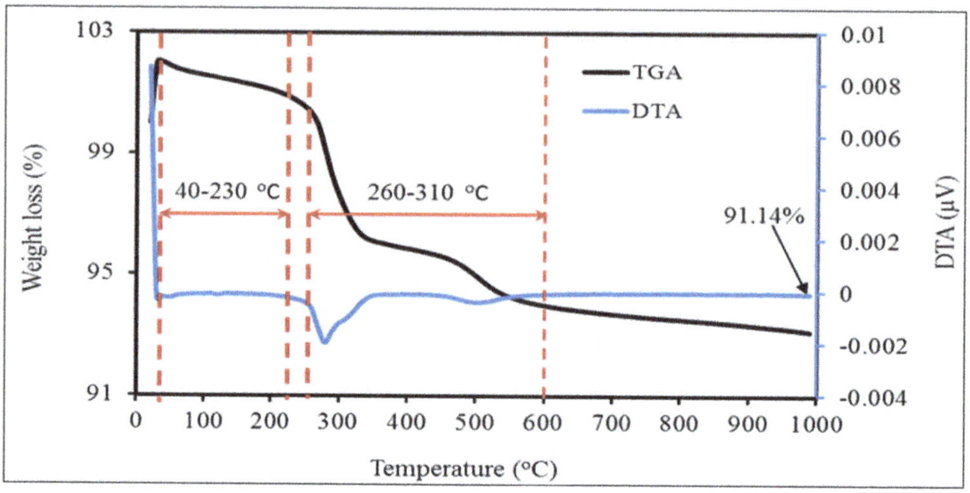

X- Ray Diffraction Pattern

The X-ray diffraction patterns of IOT are shown in Figure 2. The XRD patterns show several peaks between 12.2° and 64.1° indicating crystalline structure. The observed sharp crystalline peaks are quartz (SiO_2), gibbsite [$Al(OH)_3$], hematite (Fe_2O_3) and chamosite [$(Fe^{2+}, Mg)_5Al(AlSi_3O_{10})(OH)_8$]. Quartz is the major phase and is generally known to be unreactive (Yao et al., 2009), but usually increase the packing density and long-term strength development

Figure 2. XRD pattern of IOT

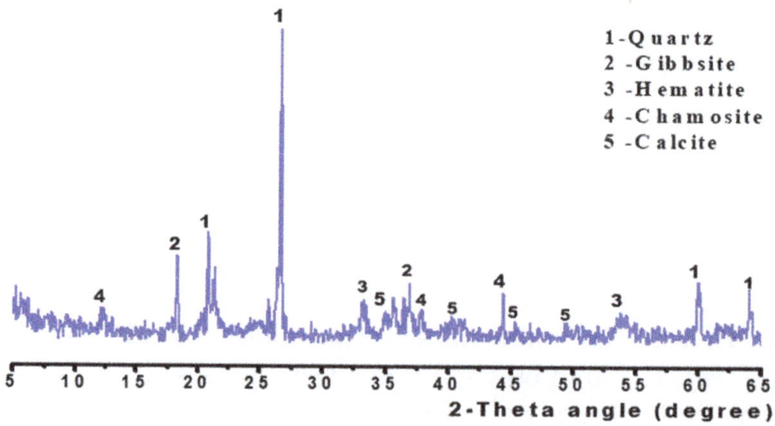

(Huang et al., 2013a). Traces of calcite ($CaCO_3$) were also detected in the spectrum and these can be attributed to the high loss on ignition observed in the chemical composition (Yao et al., 2015).

Fourier Transform Infrared Spectroscopy (FTIR)

Figure 3 presents the FTIR spectra for IOT. Results showed that the spectrum of IOT minerals have two groups of absorption; outer hydroxyl groups and inner hydroxyl group. The outer groups are in the outer unshared plane (upper), whereas the inner groups are situated in the lower shared plane of the octahedral sheet. The spectrum of the minerals was arranged in their sheets, as well as in the occupancy of the octahedral and tetrahedral sites. The inner hydroxyl groups, lying between the tetrahedral and octahedral sheets, give the absorption near 3693 cm^{-1}. A strong band at 3734 cm^{-1} is related to the in-phase symmetric stretching vibration, two weak absorptions at 3448 and 3618 cm^{-1} are assigned to out-of-plane stretching vibration (Kristof et al., 1997). The absorption peaks at 1031 and 912 cm^{-1} show Si–O stretching vibration for symmetric and asymmetric bond vibration, respectively (Kaufhold et al., 2012). The bending vibrations at 797 and 470 cm^{-1} were assigned to Si–O–Al and Si–O–Si respectively. The spectrum band at 694 cm^{-1} exhibits the Fe_2O_3 stretching vibration (Kaufhold et al., 2012), which is a characteristic of IOT minerals as shown in XRF results. The study of IOT spectra further

Figure 3. FTIR Spector of IOT

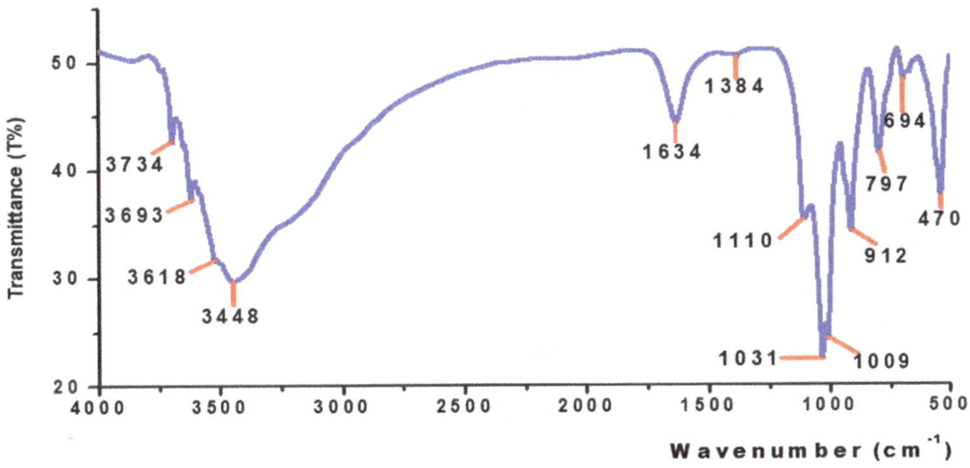

shows that the minerals' absorption bands play an important role frequently in the water absorption rate and differentiation of mineral composition due to structural OH and Si–O groups.

Field Emission Scanning Eectron Microscopy (FESEM/EDX)

The results of FESEM/EDX presented in Figure 4 showed that IOT material is of angular shape. It also consisted of porous and highly rough surface with irregular shape that was well dispersed. The material is made up of loose particles with high surface area. The nature of the particles and surface texture of IOT could contribute positively to the water demand and consequently lead to low workability. Similar behaviour has been reported by Zhao et al. (2014). Similarly, the EDX shows that the prominent element present in the materials was silica (Si) with a relative percentage by weight of 57.43%. The other elements that were more prominent were iron (Fe) and alumina (Al) comprising about 8.01% and 5.91% respectively. The alumina element plays a very important role in the formation of calcium alumina silicate hydrate (C-A-S-H) during cement-pozzolanic hydration process. The high carbon has been acknowledged as the basic reason for its high LOI value and consequently high water demand. Generally, there is a very good agreement between the findings about the properties of IOT carried out using XRD, XRF and EDX methods.

Figure 4. Field emission scanning electron microscopy morphology and energy-dispersive x-ray analysis of IOT

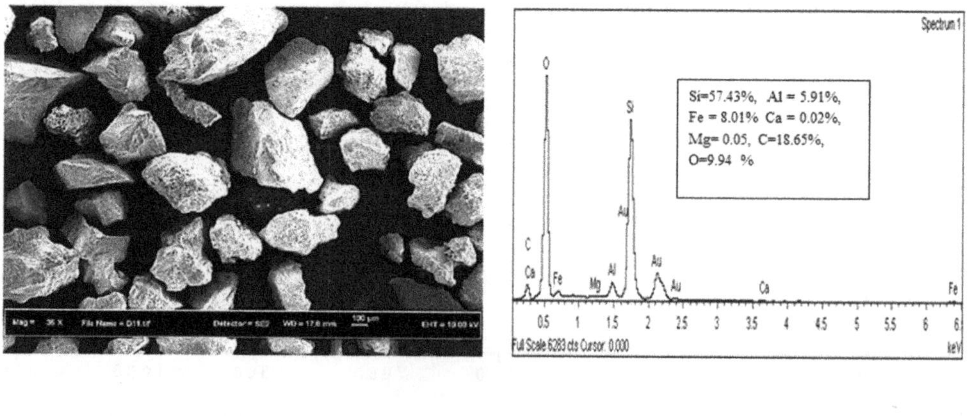

CONCLUSION

The concentration values of silica and alumina contents obtained from chemical analysis correlate with the mineralogical composition of IOT. The leaching behaviour of IOT is within the acceptable limits. From the TGA-DTA thermo-grams, the TGA curve showed that the sample decomposed from 100% to 91.14%, which correspond to the loss of 8.86%. The morphology results indicated that the IOT were porous, angular and have rough surface with irregular shape and loose particles with high surface area.

ACKNOWLEDGMENT

We, the editors greatly appreciate the exceptional contribution of Dr. Ali Umara Shettima for carrying out this research and writing of the chapter. Indeed, members of his team, Dr. Kharunisa Muthusamy, Dr. Mohd Azreen Mohd Ariffin, Dr. Mostafa Samadi, Dr. Han Seung Lee, and Dr. Muhammad Ekhlasur Rahman deserve to be equally acknowledged for their co-authorship and immense collaboration in conducting/writing of the research.

REFERENCES

Cai, J. W., Wu, J. X., Lü, Z. H., & Gao, G. L. (2011). Effects of powdery mill tailings from magnetite on workability and strength of concretes. *Key Engineering Materials*, *477*, 233–238. doi:10.4028/www.scientific.net/KEM.477.233

Cai, L., Ma, B., Li, X., Lv, Y., Liu, Z., & Jian, S. (2016). Mechanical and hydration characteristics of autoclaved aerated concrete (aac) containing iron-tailings: Effect of content and fineness. *Construction & Building Materials*, *128*, 361–372. doi:10.1016/j.conbuildmat.2016.10.031

Carrasco, E. V. M., Magalhaes, M. D. C., Santos, W. J. D., Alves, R. C., & Mantilla, J. N. R. (2017). Characterization of mortars with iron ore tailings using destructive and nondestructive tests. *Construction & Building Materials*, *131*, 31–38. doi:10.1016/j.conbuildmat.2016.11.065

Choi, Y. W., Kim, Y. J., Choi, O., Lee, K. M., & Lachemi, M. (2009). Utilization of tailings from tungsten mine waste as a substitution material for cement. *Construction & Building Materials, 23*(7), 2481–2486. doi:10.1016/j.conbuildmat.2009.02.006

Committee on Technologies for the Mining Industry. (2002). *Evolutionary and Revolutionary Technologies for Mining*. Washington, DC: National Academy of Sciences.

Coruh, S., & Ergun, O. N. (2006). Leaching characteristics of copper flotation waste before and after vitrification. *Journal of Environmental Management, 81*(4), 333–338. doi:10.1016/j.jenvman.2005.11.006 PMID:16730115

Da Silva, F. L., Da Silva Araújo, F. G., Gonçalves Castro, C., & Von Krüger, F. L. (2015). Results of the leaching, water absorption and mold release for concrete blocks with the addition of the concentration tailings of iron ore. *Materials Science Forum, 820*, 549–552. doi:10.4028/www.scientific.net/MSF.820.549

Edraki, M., Baumgartl, T., Manlapig, E., Bradshaw, D., Franks, D. M., & Moran, C. J. (2014). Designing mine tailings for better environmental, social and economic outcomes: A review of alternative approaches. *Journal of Cleaner Production, 84*, 411–420. doi:10.1016/j.jclepro.2014.04.079

Gite, B. E., Rathi, M. K., Rajguru, S. R., & Shaikh, A. P. (2013). *Advance construction material- micro silica in concrete*. Retrieved from https://www.engineeringcivil.com/advance-construction-material-micro-silica-in-concrete.html

Goyal, S., Singh, K., Hussain, A., & Singh, P. R. (2015). Study on partial replacement of sand with iron ore tailing on compressive strength of concrete. *International Journal of Research in Engineering & Advanced Technology, 3*(2), 243–248.

Grangeia, C., Ávila, P., Matias, M., & da Silva, E. F. (2011). Mine tailings integrated investigations: The case of Rio tailings (Panasqueira Mine, Central Portugal). *Engineering Geology, 123*(4), 359–372. doi:10.1016/j.enggeo.2011.10.001

Hashem, F. S., Amin, M. S., & Hekal, E. E. (2011). Stabilization of Cu (II) wastes by C3S hydrated matrix. *Construction & Building Materials, 25*(8), 3278–3282. doi:10.1016/j.conbuildmat.2011.03.015

Hitch, M., Ballantyne, S. M., & Hindle, S. R. (2010). Revaluing mine waste rock for carbon capture and storage. *International Journal of Mining, Reclamation and Environment*, *24*(1), 64–79. doi:10.1080/17480930902843102

Huang, X., Ranade, R., & Li, V. C. (2013a). Feasibility study of developing green ecc using iron ore tailings powder as cement replacement. *Journal of Materials in Civil Engineering*, *25*(7), 923–931. doi:10.1061/(ASCE)MT.1943-5533.0000674

Huang, X., Ranade, R., Ni, W., & Li, V. C. (2013b). Development of green engineered cementitious composites using iron ore tailings as aggregates. *Construction & Building Materials*, *44*, 757–764. doi:10.1016/j.conbuildmat.2013.03.088

Kaufhold, S., Hein, M., Dohrmann, R., & Ufer, K. (2012). Quantification of the mineralogical composition of clays using FTIR spectroscopy. *Vibrational Spectroscopy*, *59*, 29–39. doi:10.1016/j.vibspec.2011.12.012

Kristof, J., Frost, R. L., Felinger, A., & Mink, J. (1997). FTIR spectroscopic study of intercalated kaolinite. *Journal of Molecular Structure*, *411*, 119–122. doi:10.1016/S0022-2860(96)09488-4

Kuranchie, F. A., Shukla, S. K., Habibi, D., Mohyeddin, A., & Puppala, A. J. (2015). Utilisation of iron ore tailings as aggregates in concrete. *Cogent Engineering*, *2*(1), 1083137. doi:10.1080/23311916.2015.1083137

Maifala, B., & Tabbiruka, N. M. S. (2007). Chemical and thermal characterization of a clayey material found near Gaborone Dam. *Applied Science Environmental Management*, *11*, 77–80.

McKinnon, E. (2002). The environmental effects of mining waste disposal at Lihir Gold Mine, Papua New Guinea. *Journal of Rural and Remote Environmental Health*, *1*, 40–50.

Skarżyńska, K. M. (1995). Reuse of coal mining wastes in civil engineering—Part 2: Utilization of minestone. *Waste Management (New York, N.Y.)*, *15*(2), 83–126. doi:10.1016/0956-053X(95)00008-N

Temuujin, J., Okada, K., MacKenzie, K. J. D., & Jadambaa, T. (1999). The effect of water vapour atmospheres on the thermal transformation of kaolinite investigated by XRD, FTIR and solid state MAS NMR. *Journal of the European Ceramic Society*, *19*(1), 105–112. doi:10.1016/S0955-2219(98)00170-8

U.S EPA test method 1311. (1999). *Toxicity Characteristic Leaching Procedure*. Washington, DC: United State Environmental Protection Agency.

U.S. Geological Survey (USGS). (2017). Report on Global Iron Ore Production Data; Clarification. *Mining Engineering*.

Wills, B. A., & Napier-Munn, T. (2005). 16 - Tailings disposal. In B. A. W. Napier-Munn (Ed.), *Wills' Mineral Processing Technology* (7th ed.; pp. 400–408). Oxford, UK: Butterworth-Heinemann. doi:10.1016/B978-075064450-1/50018-7

Yao, R., Liao, S., Dai, C., Liu, Y., Chen, X., & Zheng, F. (2015). Preparation and characterization of novel glass–ceramic tile with microwave absorption properties from iron ore tailings. *Journal of Magnetism and Magnetic Materials*, *378*, 367–375. doi:10.1016/j.jmmm.2014.11.066

Yao, X., Zhang, Z., Zhu, H., & Chen, Y. (2009). Geopolymerization process of alkali–metakaolinite characterized by isothermal calorimetry. *Thermochimica Acta*, *493*(1-2), 49–54. doi:10.1016/j.tca.2009.04.002

Ye, J., Zhang, W., & Shi, D. (2014). Effect of elevated temperature on the properties of geopolymer synthesized from calcined ore-dressing tailing of bauxite and ground-granulated blast furnace slag. *Construction & Building Materials*, *69*, 41–48. doi:10.1016/j.conbuildmat.2014.07.002

Yellishetty, M., Karpe, V., Reddy, E. H., Subhash, K. N., & Ranjith, P. G. (2008). Reuse of iron ore mineral wastes in civil engineering constructions: A case study. *Resources, Conservation and Recycling*, *52*(11), 1283–1289. doi:10.1016/j.resconrec.2008.07.007

Zhang, S., Xue, X., Liu, X., Duan, P., Yang, H., Jiang, T., ... Liu, R. (2006). Current situation and comprehensive utilization of iron ore tailing resources. *Journal of Mining Science*, *42*(4), 403–408. doi:10.100710913-006-0069-9

Zhao, S., Fan, J., & Sun, W. (2014). Utilization of iron ore tailings as fine aggregate in ultra-high performance concrete. *Construction & Building Materials*, *50*, 540–548. doi:10.1016/j.conbuildmat.2013.10.019

Chapter 2
Effect of High-Volume Oil Palm Biomass Waste in Mortar

ABSTRACT

This chapter discusses the utilization of wastes in the form of palm oil fuel ash, oil palm kernel shell, and oil palm fibre in the production of mortar mixes as a part of new and innovative materials in construction industry. Detailed introduction is provided followed by a clear description of each waste and its effect when added to mortar mixes. Furthermore, a research study on the effect of palm oil fuel ash, palm oil kernel shell, palm oil fibre on mortar properties was carried out and the experimental program details are given under four subtitles. Splitting tensile strength and flexural strength were performed to test the engineering properties of mortar containing different types of waste. Results and discussion are provided for additional grasp. It is concluded that the inclusion of high-volume palm oil biomass waste can produce sustainable mortars with high strength and with more durability performance.

INTRODUCTION

When properly designed and manufactured, concrete has become a predominant building material in the construction industry due to its excellent mechanical and physical properties. More than 10 billion tonnes of concrete are produced annually. As it has been estimated that by the year 2050, the rate of world's population will increase from 1.5 to 9 billion, directly increasing the demand

DOI: 10.4018/978-1-5225-8325-7.ch002

for housing and concrete materials which are estimated to be 18 billion tonnes by 2050 (Meyer, 2009). However, with the large quantities of concrete being produced, there are consequences that will affect the environment. It has been found that approximately one tonne of carbon dioxide (CO_2) is released with every tonne of cement produced. Overall, around 5% of total CO_2 emission all over the world is contributed by the cement manufacturing industry (Worrell et al., 2001). Due to the implementation of the Kyoto protocol in February 2005, countries all over the world must reduce their greenhouse gas emissions (GHG). Therefore, the major challenges are to reduce the CO_2 emissions arising from cement manufacturing industries.

Previously, researchers have focused on finding materials that have similar properties as cement are yet more sustainable and affordable. Since cement is the major building materials, finding and substituting a material that would reduce the overall cost is necessary. The need towards sustainability and sustainable environment has made the use of pozzolanic material in construction popular. Within the past few years, natural and industrial wastes have been used in construction materials namely, pulverized fuel ash (PFA), ceramic tile waste, ground granulated blast furnace slag (GGBFS), metakaolin (MK), silica fume (SF), palm oil fuel ash (POFA) and rice husk ash (RHA), etc. Their usage as construction materials have become quite prevalent since they offer advantages in terms of strength and durability (Meyer, 2009; Samadi et al., 2015; Ismail et al., 2013; Ramadhansyah et al., 2012). In addition, these pozzolanic materials have also been proven to improve the properties of concrete and reduce the hydration temperature and hence make massive construction easier.

Palm oil fuel ash (POFA) is generally known as one of the pozzolanic materials from natural waste. By 2014, Malaysia had recorded a staggering 5.39 million hectares of oil palm plantations, an increase of 11.0% from the previous 4.85 million hectares in 2010 (MPOB, 2015). Malaysia is the second largest producer of crude palm oil in the world, consequently producing a large amount of oil palm waste. It is estimated that "the total potential palm biomass from 4.69 million hectares of oil palm planted area in Malaysia in 2009 is 77.24 million tonnes per year comprising of 13.0 million tonnes of oil palm trunks, 47.7 million tonnes of oil palm fronds, 6.7 million tonnes of empty fruit bunches, 4.0 million tonnes of palm kernel shell and 7.1 million tonnes of mesocarp fibre (all dry weight)" (Ng et al., 2011; Zwart, 2013). These wastes are usually used as fuel in oil palm mill to generate electricity and produced approximately 5% another solid waste namely, POFA (Tay and Show, 1995). In Malaysia alone, about 3 million tonnes of ashes are generated annually (MPOB, 2015). Even though the wastes have been used for other purpose, they are still left and dumped in open field creating environmental and health hazards. Therefore, the utilization of oil palm biomass waste presents an opportunity to make the industry more environmentally safe and sustainable.

Problem Statement

The rapid developments in the construction industry in Malaysia have led to huge demand for Portland cement and natural aggregates for construction. However, the use of these materials not only poses environmental risks but also depletes the natural resources as well. Therefore, this study was undertaken to find a way to reduce the use of Portland cement and natural aggregates by replacing them with high volume oil palm biomass waste namely palm oil fuel ash (POFA), oil palm kernel shell (OPKS) and oil palm fibres (OPF). Furthermore, the brittle characteristic of concrete or mortar requires another material to be added to increase the performance life of concrete. Increasing the use of waste materials reduces the use of cement and natural aggregates automatically. Large amount of oil palm biomass waste, which is locally available in large quantities, could be utilized and would solve disposal problem effectively.

Oil palm biomass waste is recommended for use in order to reduce the CO_2 emission and to recycle the waste materials for the sake of environment. Despite the application of POFA as a pozzolanic material in mortar up to 30% as cement replacement, the problem of ash disposal remains as large amount of the ash goes unutilized. Therefore, by replacing cement with high volume Nano POFA will improve the strength and durability characteristic of mortar. On the other hand, the addition of OPKS will reduce the density of high filler loading mortar while the addition of OPF solves the issues of brittle characteristic of mortar. In order to completely evaluate the potential of oil palm biomass wastes for new applications, a comprehensive and detailed study of fundamental properties is truly necessary.

Oil Palm Biomass Waste

In the oil palm mill, palm oil consists only 10% from the total biomass. The rest, 90% biomass are disposed as wastes (Awalludin et al., 2015). At present, Malaysia has at least 417 productive palm oil extraction mills nationwide

(Umar et al., 2013). Khalil et al. (2010) reported that the combination of wastes from all the mills can generate more than 12.4 million tonnes of Empty Fruit Bunch (EFB) as solid waste per year and most of them have been returned to the plantation site. In a yearly basis, Oil Palm Fibre (OPF) and Oil Palm Trunk (OPT) can mountain up to 4 and 18 million tonnes (dry weight), respectively. These numbers are based on estimation of 5% oil palm plantation that are due for new replantation (Kong et al., 2014). In 2012, an estimated 83 million tonnes (dry weight) of oil palm biomass wastes were available throughout the country and it is projected that these wastes will keep rising to reach 100 million tonnes dry weight by 2020 (Umar et al., 2013). The high potential value of these wastes to be used for more profitable purposes is often being ignored.

Lack of technology to utilize the entire remaining wastes can worsen the biomass overload problem for the environment. However, the production of Oil Palm biomass waste is limited to only a few countries such as Thailand, Indonesia, Malaysia, Nigeria, etc. Therefore, it is not widely used in the construction projects. Fewer case studies have been reported by using oil palm biomass waste in the application of concrete in construction works. In most of these countries, the oil palm biomass waste was used to producc full scale concrete beams for the structural tests in the laboratory (Alengaram et al., 2011; Aldahdooh et al., 2014).

In Malaysia, the oil palm shells have been utilized as replacement of aggregates to produce low cost houses in Sabah (Teo et al., 2006) and POFA as cement replacement in the production of eco-homes in Johor (Ahmad, 2016). The POFA used in this study was black in colour. It was obtained from palm oil mill located in Johor which is southern part of Malaysia. The POFA was treated and ground until nanoparticles sizes. The POFA used was similar to the studies done by Lim et al. (2015). The chemical composition of OPC and nano POFA (NPOFA) used in this study are shown in Table 1.

The collected OPKS was subjected to further grinding to obtain similar properties as sand. The use of OPKS was to reduce the density of high-volume palm oil biomass mortar. The bulk density of OPKS and sand used were 750 kg/m^3 and 1614 kg/m^3, respectively. The size used was in the range of 300 μm – 2.35 mm. The palm oil fibres (OPF) used in this research was obtained from the empty fruit branch of oil palm tree. After oil extraction process, the fruits or nuts were stripped from fruit bunches, leaving behind the empty – fruit bunches as waste. The valuable fibre was obtained from the oil palm empty fruit branches after it had been treated in boiled water for 3 hours at 100 ±5°C.

Table 1. Chemical composition of OPC and NPOFA

Chemical Composition (%)	OPC	NPOFA
SiO_2	16.40	69.30
Al_2O_3	4.24	5.30
Fe_2O_3	3.53	5.10
CaO	68.30	9.15
K_2O	0.22	11.10
MgO	2.39	4.10
CO_2	0.10	0.10
SO_3	4.39	1.59
LOI	2.40	1.30

All mortar specimens were prepared with blended ash to fine aggregate ratio of 1:3, whereby the fine aggregate was prepared in ambient air-dry condition. The percentages of fibre were varied from 0.3 to 0.9%. The mixing was carried out at a room temperature of approximately 28°C. The mix proportions for all mixes are given in Table 2. The test specimens of 70 x 70 x 70 mm cube, 40 x 40 x 160 mm prism and 70 Φ x 150 mm cylinder were prepared. The specimens were compacted in two-layer with tamping as described in ASTM C109-11. The specimens were demoulded after 24 hours and placed in water tank for curing process until the age of testing for compressive strength test, flexural strength test and splitting tensile strength test.

Effect of Palm Oil Fuel Ash

The effect of NPOFA replacement on strength performance of mortar is shown in Figure 1. The 60% NPOFA replacement shows higher strength at

Table 2. Mix proportions of biomass mortars

Materials (kg/m³)	Mortar Mix			
	OPC	60% POFA	80% POFA	100% POFA
OPC	525	210	105	-
POFA	-	315	420	525
Sand	1578	1578	1578	1578
OPKS	0, 25, 50, 75, 100%			
w/c ratio	0.4	0.4	0.4	0.4

Figure 1. Effect of percentage of NPOFA on compressive strength

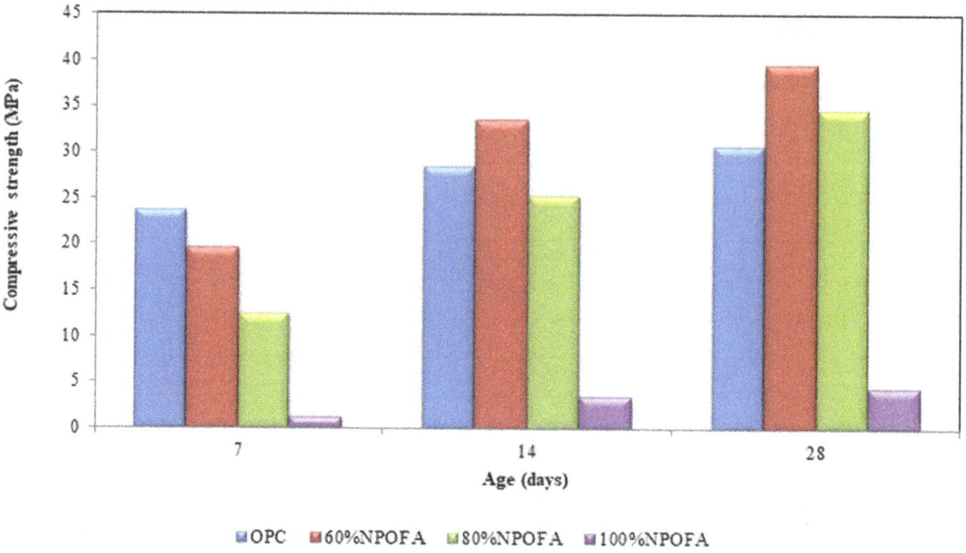

all curing age compared with the other mixes. This is due to the appropriate amount of reactive silica for pozzolanic reaction and $Ca(OH)_2$ from the cement hydration process. However, as the percentage of NPOFA increases, the compressive strength of mortar decreases. This is probably due to the lack of calcium hydroxide to react with the reactive silica from NPOFA resulting in less calcium silicate hydrate gel (C-S-H). The compressive strength of mortar increases as the age increases due to the hydration process that causes an increase in the C-S-H gel. In addition, it is believed that the finer particles will fill the voids inside the mortar and make it denser than normal OPC mortar. Besides, the compressive strength at 28 days for 90% and 100% replacement are reduced by 73% and 86%, respectively. This is probably due to the lack of CaO from binder to start the hydration process thus, produce less $Ca(OH)_2$ as the cement content reduced.

Previous researches suggested that the strength of mortar with pozzolanic materials is expected to be higher at later ages due to its pozzolanic reaction activities that resembles with other blended cement mortar behaviour (Awal and Shehu, 2011). However, by using nano size of POFA, the early strength also increased. This is due to the reaction by K_2O content in NPOFA which is higher (11.10%) that initiates the early strength of mortar. Mortar samples made with higher potassium (K_2O) content accelerated the formation of ettringite thus increased the early strength. This was also highlighted by Cohen and

Richards (1982) who mentioned that the formation of ettringite was enhanced with the presence of alkalis. Overall, by increasing the amount of NPOFA, the compressive strength of mortar increased until a certain limit where the compressive strength reduced drastically compared with normal mortar.

Effect of Oil Palm Kernel Shell

The effects of OPKS percentages on compressive strength of mortar are shown in Figure 2. Cement was substituted by 80% of NPOFA in all the mortar mixes. Sand was replaced with 25%, 50%, 75% and 100% of OPKS. The recorded compressive strength at age of 28 days for 0%, 25%, 50%, 75% and 100% OPKS replacement were 34.73, 32.89, 30.65, 23.65 and 18.76 MPa, respectively. Evidently, the compressive strength was decreased as the OPKS percentage increased. This is due to the mechanical properties of OPKS, which was derived from oil palm biomass waste as reported by Okpala (1990) and Basri et al. (1999). The compressive strength of mortar was affected by the strength, thickness and density of the OPKS aggregates. The OPKS aggregates have flaky shape which reduced the bonding between fine aggregates and cement matrix thus causing a reduction in the strength of mortar.

Mannan and Ganapathy (2004) reported that at the early age, the failure happened due to the breakdown between binder and fine aggregates, while at

Figure 2. Effect of OPKS on compressive strength of mortar

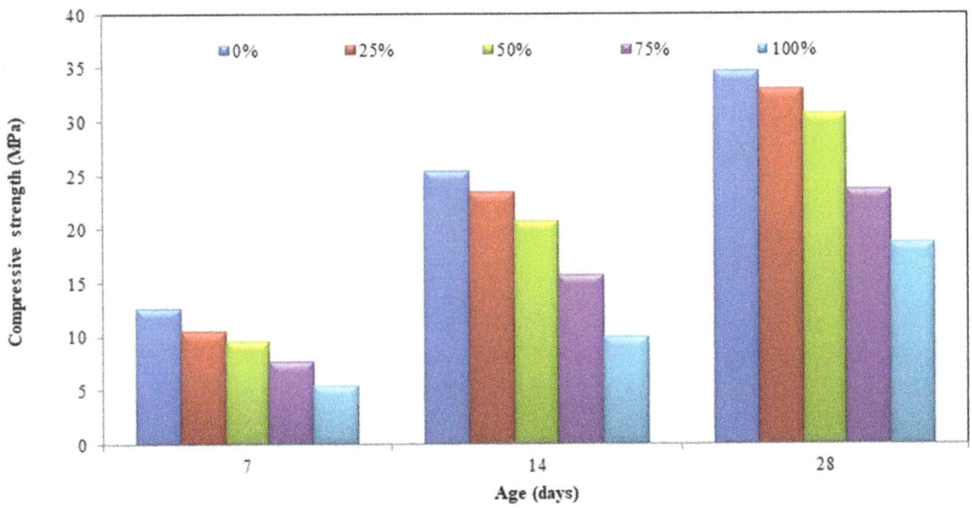

later age the failure happened due to the breaking of the OPKS itself. However, in this study since the size of OPKS is reduced, the failure happened due to the breakdown of the binder and fine aggregates. The aggregates strength and binder strength were found to be an important factor in reducing the compressive strength. However, with the addition of 50% OPKS at 28 days shows relatively similar value to the target strength which is 30.7 MPa. The flakiness shape of OPKS significantly decreased by crushing the larger sizes of OPKS aggregates which resulted in a better performance as fine aggregates and consequently offered better compressive strength.

Effect of Oil Palm Fibre

The compressive strength of oil palm biomass waste with different OPF ratios is shown in Figure 3. The compressive strength for all mortar mixes showed similar strength development with increasing age. It was found that the fibres did not affect the compressive strength of mortar as much. The compressive strength at 28 days for 0, 0.3, 0.5, 0.7 and 0.9% OPF were 34.7, 35.1, 35.7, 36.9 and 36.2 MPa, respectively. The compressive strength of mortar decreased with the addition of 0.9% OPF. This may be due to a higher

Figure 3. Effect of OPF percentage on compressive strength of biomass mortar

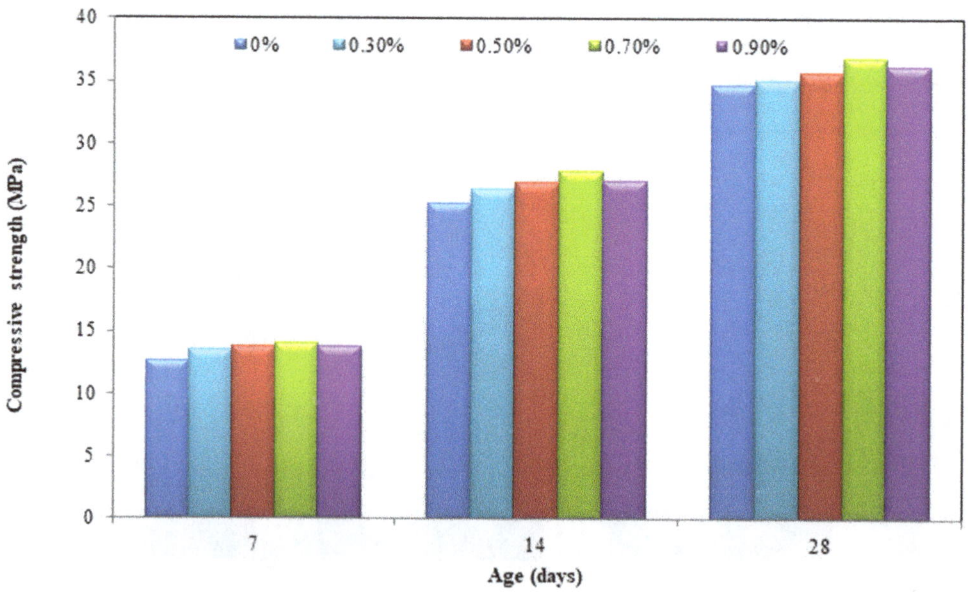

void volume content within the fibres themselves, which leaves room for more compressibility of the samples and results in reducing the compressive strength. Moreover, due to the addition of higher volume of OPF, the fibres tend to agglomerate and cause a balling effect. By adding 0.7% OPF, the compressive strength of mortar at 28 days increased by 6% compared with normal mortar. This result indicated a bond strength improvement with the addition of OPF that interconnected the matrix in the composites and reduced the crack visibility in mortar. This shows a good bonding can be developed with the addition of 0.7% fibres in biomass mortar.

Splitting Tensile Strength Test

The splitting tensile strength of biomass mortar with different percentages of OPF is shown in Figure 4. The splitting tensile strength values at the age of 28 days were 1.2, 1.9, 2.5, 4.0 and 2.8 MPa for mortars with 0, 0.3, 0.5 0.7 and 0.9% fibres addition, respectively. The splitting tensile strength of biomass mortar found to be increased with the increasing ages of curing. The 0.7% OPF addition showed the highest splitting tensile strength at all ages. This is due to the function of the fibre that helps to carry load and withstand the stress after cracking thereby increasing the tensile strength.

Figure 4. Splitting tensile strength of biomass mortar with different percentage OPF

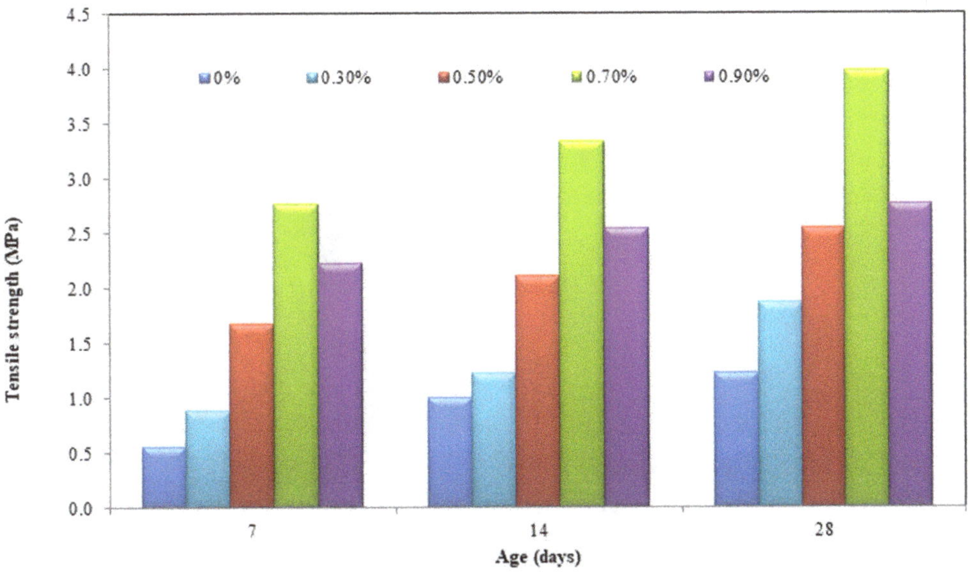

During the splitting tensile test, the effect of the fibres was apparent. The fibres appeared to control the cracking and alter the post cracking behaviour. The fibres seem to provide a load redistribution mechanism after initial cracking. However, it was observed that with the inclusion of 0.9% fibres, the splitting tensile strength of the mortar was decreased. This is probably due to the high-volume inclusion of fibres that tends to agglomerate and causes balling effect. Similar findings were reported by previous researchers where with the inclusion of more than 1% fibres, the tensile strength reduced (Dawood and Ramli, 2011).

Figure 5 shows the typical failure patterns of mortar without fibres and with the addition of 0.7% OPF after splitting tensile tests. It is shown that the mortar specimens without fibres split into two main halves along the loading path as expected while mortar specimen with 0.7% fibres still maintain the original shape. Unlike normal mortar, it was difficult to separate the fractured specimens because the fibres were bridging the gap that kept the two mortar sections together as shown in Figure 5 (b).

Flexural Strength Test

Figure 6 shows the flexural strength test for biomass mortar with different percentages of OPF. The flexural strength values at the age of 28 days were 2.6, 3.5, 4.8, 7.2 and 5.3 MPa for mortars with 0, 0.3, 0.5 0.7 and 0.9% fibres,

Figure 5. Failure pattern of mortar after splitting tensile strength test

a) mortar without fibre b) mortar with 0.7% fibre

Figure 6. Flexural strength of biomass mortar with different percentage of OPF

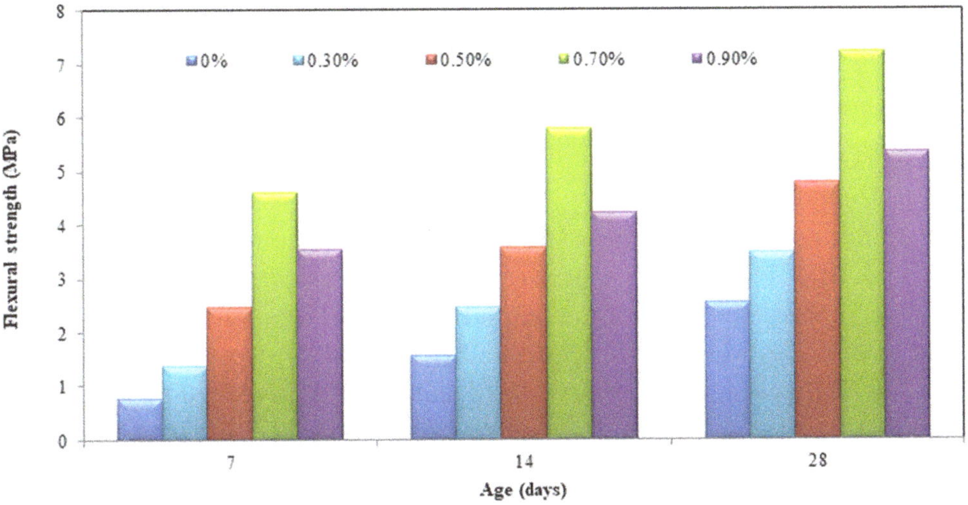

respectively. Similar to the splitting tensile strength, the 0.7% OPF ratio showed the highest flexural strength compared with the others. This is due to the fibres themselves which played an important role in mortar by providing good fibre-matrix bonding, and ability to interrupt the distribution of stress that resulted in multi cracking in mortar. The flexural strength for all ages of curing decreased with the addition of 0.9% OPF. This could be attributed to the balling effect of fibres taking place during mixing. In addition, the reduction in flexural strength was attributed to the agglomeration of the filler and its inability to support stresses transferred from the binder matrix. Furthermore, with the high volume of fibres, it contributed to weaker bonding between particles. In addition, the flexural strength continued to increase as the hydration period and pozzolanic reaction increased.

CONCLUSION

Based on the experimental investigation, the following remarks can be highlighted.

The use of Nano size POFA increases the compressive strength of mortars even at early ages of curing. Besides, 50% OPKS can be used as fine aggregates replacement to produce lighter mortar with comparable strength. The 0.7% ratio of OPF increases the splitting tensile strength and flexural strength of biomass mortar by 69% and 65%, respectively, at age of 28 days. The oil palm biomass waste can potentially be used up to 80% replacement of cement and fine aggregates with comparable strength (30 MPa).

ACKNOWLEDGMENT

We, the editors greatly appreciate the exceptional contribution of Dr. Nor Hasanah Abdul Shukor Lim for carrying out this research and writing of the chapter. Indeed, members of her team, Dr. Mostafa Samadi, Dr. Nur Hafizah Abd Khalid, Dr. Mohd Azreen Mohd Ariffin, Dr. Ali Umara Shettima, and Dr. Nur Farhayu Ariffin deserve to be equally acknowledged for their co-authorship and immense collaboration in conducting/writing of the research.

REFERENCES

Ahmad, N. (2016, October 10). UTM team builds energy-saving home. *The Star (Online)*. Retrieved from http://www.thestar.com.my/news/nation/2016/12/12/utm-team-builds-energysaving-home

Aldahdooh, M. A. A., Bunnori, M. N., & Johari, M. A. M. (2014). Influence of palm oil fuel ash on ultimate flexural and uniaxial tensile strength of green ultra-high performance fibre reinforced cementitious composites. *Materials & Design*, *54*, 694–701. doi:10.1016/j.matdes.2013.08.094

Alengaram, J. U., Jumaat, M. Z., Mahmud, H., & Fayyadh, M. M. (2011). Shear behaviour of reinforced palm kernel shell concrete beams. *Construction & Building Materials*, *25*(6), 2918–2927. doi:10.1016/j.conbuildmat.2010.12.032

Awal, A. S. M. A., & Shehu, I. A. (2011). Properties of concrete containing high volume palm oil fuel ash: A short-term investigation. *Malaysian Journal of Civil Engineering*, *23*(2), 54–66.

Awalludin, M. F., Sulaiman, O., Hashim, R., & Nadhari, W. N. A. W. (2015). An overview of the oil palm industry in Malaysia and its waste utilization through thermochemical conversion, specifically via liquefaction. *Renewable & Sustainable Energy Reviews*, *50*, 1469–1484. doi:10.1016/j.rser.2015.05.085

Basri, H. B., Mannan, A., & Zain, M. F. M. (1999). Concrete using waste oil palm shells as aggregate. *Cement and Concrete Research*, *29*(4), 619–622. doi:10.1016/S0008-8846(98)00233-6

Cohen, M. D., & Richards, C. W. (1982). Effects of the particle sizes of expansive clinker on strength-expansion characteristics of type K expansive cements. *Cement and Concrete Research*, *12*(6), 717–725. doi:10.1016/0008-8846(82)90034-5

Dawood, E. T., & Ramli, M. (2011). Properties of high strength flowable mortar reinforced with different fibers. *Concrete Research Letters*, *2*(4), 315–325.

Ismail, M., Yusuf, T. O., Noruzman, A. H., & Hassan, I. O. (2013). Early strength characteristics of palm oil fuel ash and metakaolin blended geopolymer mortar. *Advanced Materials Research*, *690*, 1045–1048. doi:10.4028/www.scientific.net/AMR.690-693.1045

Khalil, H. P. S. A., & Fazita, N. (2010). Development and material properties of new hybrid plywood from oil palm biomass. *Materials & Design*, *31*(1), 417–424. doi:10.1016/j.matdes.2009.05.040

Kong, S. H., Loh, S. K., Bachmann, R. T. A., Rahim, S., & Salimon, J. (2014). Biochar from oil palm biomass: A review of its potential and challenges. *Renewable & Sustainable Energy Reviews*, *39*, 729–739. doi:10.1016/j.rser.2014.07.107

Lim, N. H. A. S., Ismail, M. A., Lee, H. S., Hussin, M. W., Sam, A. R. M., & Samadi, M. (2015). The effects of high volume nano palm oil fuel ash on microstructure properties and hydration temperature of mortar. *Construction & Building Materials*, *93*, 29–34. doi:10.1016/j.conbuildmat.2015.05.107

Malaysian Palm Oil Board (MPOB). (n.d.). *Economic and industry development division. Oil Palm Planted Area*. Retrieved December 1, 2015, from http://bepi.mpob.gov.my/index.php/statistics/area.html

Mannan, M. A., & Ganapathy, C. (2004). Concrete from an agricultural waste-oil palm shell (OPS). *Building and Environment*, *39*(4), 441–448. doi:10.1016/j.buildenv.2003.10.007

Meyer, C. (2009). The greening of the concrete industry. *Cement and Concrete Composites*, *31*(8), 601–605. doi:10.1016/j.cemconcomp.2008.12.010

Ng, F. Y., Yew, F. K., Basiron, Y., & Sundram, K. (2011). A renewable future driven with malaysian palm oil-based green technology. *Journal of Oil Palm and the Environment*, *2*, 1–7.

Okpala, D. C. (1990). Palm kernel shell as a lightweight aggregate in concrete. *Building and Environment*, *25*(4), 291–296. doi:10.1016/0360-1323(90)90002-9

Ramadhansyah, P. J., Mahyun, A. W., Salwa, M. Z. M., & Bakar, B. H., Johari, M. A. M., & Ibrahim, M. H. (2012). Thermal analysis and pozzolanic index of rice husk ash at different grinding time. *Procedia Engineering*, *50*, 101–109.

Samadi, M., Hussin, M. W., Lee, H. S., Sam, A. R. M., Ismail, M. A., Lim, N. H. A. S., Ariffin, N. F., & Khalid, N. H. (2015). Properties of mortar containing ceramic powder waste as cement replacement. *Jurnal Teknologi*, *12*(77), 93–97.

Tay, J., & Show, K. (1995). Use of ash derived from oil-palm waste incineration as a cement replacement material. *Resources, Conservation and Recycling*, *13*(1), 27–36. doi:10.1016/0921-3449(94)00012-T

Teo, D. C. L., & Mannan, M., & Kurian, V. J. (2006). Structural concrete using oil palm shell (OPS) as lightweight aggregate. *Turkish Journal of Engineering Environment Science*, *30*, 251–257.

Umar, M. S., Jennings, P., & Urmee, T. (2013). Strengthening the palm oil biomass renewable energy industry in malaysia. *Renewable Energy*, *60*, 107–115. doi:10.1016/j.renene.2013.04.010

Worrell, E., Price, L., Martin, N., Hendriks, C., & Meida, L. O. (2001). Carbon dioxide emissions from the global cement industry. *Annual Review of Energy and the Environment*, *26*(1), 303–329. doi:10.1146/annurev.energy.26.1.303

Zwart, R. (2013). Opportunities and challenges in the development of a viable malaysian palm oil biomass industry. *Journal of Oil Palm and the Environment*, *4*, 41–46.

Chapter 3
Performance of Mortar Grouts Incorporating Rice Husk Ash and Fly Ash

ABSTRACT

Chapter 3 is based on performance of mortar grouts incorporating rice husk ash (RHA) and fly ash (FA). Detailed experimental work was conducted to investigate the mechanical properties of mortar grout using RHA and FA as partial replacement of cement. This study investigated the compressive strength and durability of mortar grouts in their hardened state. Durability tests such as water absorption, apparent volume of permeable voids, sorptivity, and rapid chloride penetration tests are researched. Detailed results and discussion which focused on mechanical properties as well as durability of hardened state mortar grout are presented. It was confirmed that the inclusion of blended RHA and FA significantly improved the compressive strength of mortar grouts. The durability of mortar grout increased along with a longer curing time. Hence, RHA and FA can partially replace cement in the production of mortar grouts.

INTRODUCTION

Deterioration of concrete structures has shortened the service life of existing reinforced concrete structures (Horrigmoe, 2000). The major causes of concrete deterioration are a lack of consideration in design details, specifications, poor labour skills, etc. This leads to decrease in durability properties such as

DOI: 10.4018/978-1-5225-8325-7.ch003

carbonation and chloride-induced corrosion of reinforcing steel that affects the long-term performance of concrete structures. The consequences of a concrete structural failure are severe and can be life-threatening. Spalling or cracking of concrete cover is the most common type of deterioration and the simplest way to rectify this condition is to use mortar grouting by patching the affected area. This will result in an extended service life of the existing reinforced concrete structure (Imbin et al., 2013). Mortar grout consists of a mixture of cement and other fine material such as fine sand. With the rising awareness of sustainability and environmental impact of construction materials, agricultural waste residues such as rice husk ash (RHA), palm oil fuel ash (POFA) and bagasse ash, in addition to, industrial wastes such as fly ash (FA), silica fume and ground-granulated blast furnace slag is increasingly used as supplementary cementitious materials (Nagaratnam et al., 2016; Sanal, 2017; Sanal 2018). This is to achieve high performance, good quality and low-cost concrete mixtures (Uduweriya and De Silva, 2010).

Rice husk ash (RHA) is a by-product from the burning of rice husk. The milling of rice generates on average at 20% rate by weight of the 500 million tons of paddy produced in the world (Ganesan et al., 2008). Subsequently, the burning of these husks produces ash at an average of 18% by weight of the husks (Reddy and Alvarez, 2006). The pozzolanic activity of RHA depends on certain criteria such as the silica content of the ash, the silica crystallization phase, the size and the surface area of its particles and a low carbon content (Zain et al., 2011; Rahman et al., 2014). A case study was reported that addressed the technical considerations in Japan (Tateda et al., 2016) for the optimisation of rice husk burning in a boiler to retain a high solubility of silica in RHA. This study described step by step details to achieve the optimal operation of a boiler for stakeholders using RHA. The best quality of silica, based on a solubility evaluation, in rice husk ash and heat could be obtained simultaneously through the optimal operation. Since the boiler used in that study was very simple and inexpensive, the best practices reported at this study can be transferred and applied to using rice husks as a resource in many parts of world, especially in Southeast Asian countries].

Fly ash, on the other hand, is the non-combustible mineral waste from the coal industry. It is collected from the power generating plant's exhaust, after the carbon is consumed in the power plant boiler and solidify as microscopic or glassy spheres (Papadakis and Tsimas, 2002). Similarly, it is also pozzolanic in nature and further aids flow-ability of cement-based materials (Nagaratnam et al., 2016). The application of RHA and FA in blended mortar grout is relatively unexplored. A significant amount of these waste materials is generated every year in many developing countries. Such wastes contain around 90% SiO_2 and due to this high percentage of SiO_2, it works as a pozzolanic material. Therefore, it is very important to utilize it in any cement-based materials like mortar grout and to further examine its hardened state properties.

This detailed experimental work was conducted to investigate the mechanical properties of mortar grout using RHA and FA as partial replacement of cement. It analyses the compressive strength and durability of the mortar grout in the hardened state. The durability tests investigated are water absorption, apparent volume of permeable voids (AVPV), sorptivity and rapid chloride penetration tests (RCPT). All the experiments were carried out in triplicate and mean values are reported.

Preparation of Supplementary Cementing Materials

Rice Husk Ash (RHA)

The RHA used in this research was obtained through an open burning condition in a rice mill in Kinarut, Sabah, Malaysia. The raw RHA obtained was coarser than the specified recommendation by ASTM C618, hence, sieving and grinding was carried out to obtain the desired fineness of RHA in order to comply with the standard.

The RHA was first dried in the oven for 24 hours to remove any moisture for the ease of sieving. Dried RHA was then sieved and collected at the size of 75 μm. It was followed by grinding in a Los Angeles abrasion machine for 180 minutes.

In order to obtain RHA with high amorphous silica and large surface area, it was produced at a controlled temperature and duration. The RHA was burnt for 30 minutes, followed by air supply of 60 minutes and a chilling process of 2 days. RHA produced by this method contains low carbon (LOI of 3.21%) and a whitish colour. On the other hand, another key process in the production of a better quality RHA is the grinding of burnt ash. A grinding duration of 60 minutes or more is suggested to get the standard fineness of RHA.

Fly Ash (FA)

The FA used in this research was supplied by Gobel Industries Sdn. Bhd. and sourced from the Sejingkat Coal-fired Power Plant in Kuching, Sarawak,

Malaysia. The samples were mostly greyish in colour. ASTM C618 specifies the physical requirements and states that not more than 34% by weight of FA should be retained on a US standard 325 mesh sieves which is a 45 μm sieve size. It can be seen from the sieve analysis that the FA used in this research conforms to the requirement as stated above as only 11% was retained on the 45 μm sieve.

The soundness of FA using the Le-Chatelier apparatus showed an expansion of about 2.4 mm which is below the maximum specified value of 10 mm and is sound for the intended use.

Preparation of Mix Design

The mix design of mortar grout incorporating FA and RHA was based on various trial mix designs according to the workability conditions of a target slump height of 50 – 180 mm as seen in most structural applications (Shan, 2001). Water to binder ratio (w/b) was kept constant at 0.42 throughout this study while the SP content adjusted to maintain the mortar grout mixes with the target slump specified.

The mix proportion of mortar grout is shown in Table 1. Cement and fine aggregates were properly mixed together in the ratio of 1:2.75 by weight before the addition of water. OPC was partially replaced with pozzolans at the dosage of 0-30% by weight of the materials. This research was carried

Table 1. Mortar grout mix proportions

Mix No.	Symbol	OPC	RHA	FA	SP (%)
1	OPC	100	-	-	1.0
2	10RHA	90	10	-	1.1
3	20RHA	80	20	-	1.4
4	30RHA	70	30	-	1.9
5	5RHA5FA	90	5	5	1.2
6	10RHA10FA	80	10	10	1.5
7	15RHA15FA	70	15	15	1.8
8	10FA	90	-	10	0.9
9	20FA	80	-	20	0.8
10	30FA	70	-	30	0.4

Note for Mix No 1: OPC = 550 kg/m^3, Sand = 1008 kg/m^3, Crushed quartz aggregates = 504kg/m^3, W/B = 0.42, Slump = 50-180 mm.

out with 10%, 20% and 30% replacement of cement by RHA and FA in the ratio of 1:0, 0.5:0.5 and 0:1 in each percentage of replacement.

Mixing, Casting and Curing of Specimens

All ingredients of grout were mixed into a cylindrical pan mixer with a vertical rotational blade. Initially, fine aggregates, cement and the supplementary cementitious materials (FA and RHA) were mixed completely in the mixer. Water was added slowly within the area of the mixer followed by the addition of super-plasticiser gradually. The total mixing time was 5 minutes. After the fresh state tests of the mix had been carried out, the mortar grout was cast into the cylindrical moulds that comply with all the requirements stated in AS1012.8.1. The cylindrical mould used was 100 mm diameter 200 mm in height. A vibration machine was used for the compaction of the mortar grout during placing. According to the AS1012.8.1, the placing of mortar grout should be finished within 20 minutes after the mixing of mortar grout. De-moulding of mortar grout specimens has to be carried out after 18 – 36 hours. For the ease of de-moulding of the specimens, a layer of oil was applied onto the inner surface of cylindrical steel. Lastly, the specimen was immersed in the curing tank for the desired curing time.

Mechanical Properties Test on Harden State Mortar Grout

Hardened state of mortar grout consisted of mechanical and durability properties. For the mechanical properties, compressive strength test was carried out. Meanwhile, the immersed water absorption test, apparent volume of permeable voids test (AVPV), sorptivity test and rapid chloride ion penetration test (RCPT) were conducted for the durability properties of the mortar grout.

Compressive Strength Test

The compressive strength test of mortar grout was carried out in accordance to the AS1012.9 (Standards Australia 1999). Testing machine was the Universal Testing Machine (UTM) which complies with AS 2193 (Standards Australia 2005). The specimen was removed from the curing tank and maintained in moist condition before running the compression test. Capping was required for specimen in any of the conditions stated in clause 4 of AS1012.9.

The specimen was cleaned to remove dirt before commencing the test. The axis of the specimen was aligned with the centre of thrust of the upper platen. After that, load was applied at the loading rate of 160KN per minute until the specimen failed and the maximum load was recorded.

Durability Test on Harden State Mortar Grout

For the durability tests, the 200 mm standard cylinder specimen was sliced into 4 pieces using a concrete cutting blade into 50 mm in height and 100 mm diameter specimens. All the tests were carried out on three specimens for each curing time, which were 28 and 84 days. Water absorption and volume of permeable voids were tested according to AS1012.21, while, sorptivity test and rapid chloride ion penetration tests were carried out with ASTM C1585 and ASTM C1202 respectively.

Water Absorption Test

Immersed water absorption test was used to indicate the amount of water absorbed by the concrete specimen at a desired age. At every test age, the initial weight of specimen was recorded before oven drying in the oven at 110 °C for 24 hours. After that, the specimen was weighed again to ensure that there was no more than 1 gram of difference. The specimen would be oven dried again until the differences were not more than 1 gram.

After letting the specimen cool to a temperature between 21 °C and 25 °C, the specimen mass was recorded as M1 and immersed in water for at least 48 hours. The dry surface weight (by using a dry cloth) was recorded as M2 and the percentage of absorption was calculated by the formula:

$$Ai = \frac{\left(M_{2i} - M_1\right)}{M_1} \times 100\% \tag{1}$$

Apparent Volume of Permeable Void (AVPV) Test

The specimen from the previously immersed absorption test can be used in AVPV test since they were non-destructive. In AVPV test, the surface dry specimen was boiled in water at 95 °C for 5.5 ± 0.5 hours and was left to cool down naturally in same water for at least 14 hours. The mass of boiled specimen was recorded as M3b. The specimen was again placed back in water and apparent weight was noted as M_{4ib}. The AVPV was calculated as follows:

$$AVPV = \frac{\left(M_{3b} - M_1\right)}{\left(M_{3b} - M_{4ib}\right)} \times 100\% \tag{2}$$

Sorptivity Test

Specimen was oven dried at 50 °C for 24 hours and mass was recorded. Then, the curved surface and one of the end surfaces were coated with silicone to seal all the surfaces and air dried for 1 hour. After that, the exposed surface (a, m_t) was contacted with water where the immersion was maintained between 1mm to 3 mm. The mass of saturated specimen (dry side which the surface water was removed) was measured at a regular interval of 60s, 300s, 600s, 1200s, and 1800s, until 8th days. The specimen had to be placed back into the immersion immediately after each measurement. The absorption, I was calculated by using the formula below:

$$I = \frac{m_t}{a \times d} \tag{3}$$

where m_t = change in specimen mass, a = exposed area and d = density of water.

The initial rate of absorption was obtained from the beginning till sixth hour. The rate of absorption was analyzed based on the slope of regression lines obtained from the graph of absorption (I) against square root of time ($S^{0.5}$).

Rapid Chloride Ion Penetration (RCPT) Test

Before commencing the test, all the specimens were pre-conditioned. Initially, the curved surface of the specimen was coated with silicone and air dried for 1 hour. The specimen was then placed in a vacuum desiccator with both ends exposed. After sealing the desiccator, a vacuum was started and maintained for 3 hours. Pressure should be decreased to less than 50 mm Hg (6650 Pa) within a few minutes. With the vacuum pump still running, sufficient distilled water to cover the specimens was drained into the desiccator (air was not allowed to enter from outside). Vacuum pump was still running for additional an hour under this condition. The specimen was left soaked in water for a further 18 hours to complete the pre-conditioning procedure. After the pre-conditioning, the specimen was removed from the desiccator and then clamped together between two halves of a voltage cell. One side of the cell was filled with 3.0% of NaCl solution while the other side was filled with 0.3 N NaOH solution.

The electrical wires were connected to appropriate apparatus. The 60.0 ± 0.1V potential difference was applied across the cell for 6 hours and the current passing through the specimen was recorded as at intervals of 30 minutes until 6 hours. During the test, the temperature of the solutions was not allowed to exceed 90 °C. Since the requirement of the standard was achieved, the total charge passed was calculated directly with the formula:

$$Q = 900(I_0 + I_{30} + I_{60} + \ldots\ldots I_{360}) \quad (4)$$

RESULTS AND DISCUSSIONS

Mechanical Properties Test on Harden State Mortar Grout

Compressive Strength Test

Figure 1 shows the results of the average compressive strength tests of the mortar grout containing various percentage replacements of RHA and FA. The strength of the mortar grout containing combination of RHA and FA is relatively high. It can be observed that the mortar grout mix incorporating only FA achieved a lower compressive strength than all the control mix for all tested times. However, the addition of RHA had generally improved the strength at all ages. At all ages, the strengths of the mortar grout mixtures containing only RHA basically achieved higher strength than the control mix except for the 30RHA mixtures because the strength was found to decrease with increasing the amount of RHA.

According to Habeeb and Mahmud (2010), this could be explained in terms of remaining silica from RHA and the amount of $Ca(OH)_2$ from the hydration process increases the pozzolanic reaction. The amounts of silica available in the hydrated blended cement reacts with the available amount of produced $Ca(OH)_2$ and left no silica to act as inert material which would contribute nothing to the strength.

The higher content of RHA causes low strength due to the excessive amount of silica present in the mix. The extra amount of silica from RHA leaches out and replaces part of the cementitious material without contributing to the strength (Al-Khalaf and Yousif, 1984). Mixtures of combination of RHA and FA can be seen to generally improve the compressive strength.

Givi et al. (2010) reported that addition of RHA provides a good effect on compressive strength of concrete at early ages. However, FA tends to reduce the early age strength and only increases in the long term (Naik et al., 1995). This could explain why the strengths of all the mixtures containing RHA and FA are not much different from the strength of control mix at early ages.

The strengths of mixtures with 10%, 20% and 30% FA are lower as compared to the control mix. Low calcium fly ash tends to reduce the early age strength which retards the hydration of C_3S and thus its pozzolanic reaction is slow and can only contribute to the strength gain at later ages (Islam and Islam, 2010).

The compressive strength gain of mortar grout incorporating RHA and FA with curing time can be studied as well in Figure 1. At 28 days, the strength of mixture 5RHA5FA rose higher than that of 10RHA mixture which has the highest strength at the 7 days. Mixture 5RHA5FA continued to gain strength and remained the highest from 28 days to 84 days. This indicated that 5% inclusion of RHA and 5% of FA in mortar grout could be better in compressive strength than all the other mixtures. Besides, these results not only show high pozzolanic activity but also slow pozzolanic reaction rate in mixtures containing RHA and FA.

The compressive strength of 10FA was 37.1 MPa which was only slightly lower than the strength of control mix at 84 days. This could be explained

Figure 1. Compressive strength of mortar grout with various percentages replacement of RHA and FA

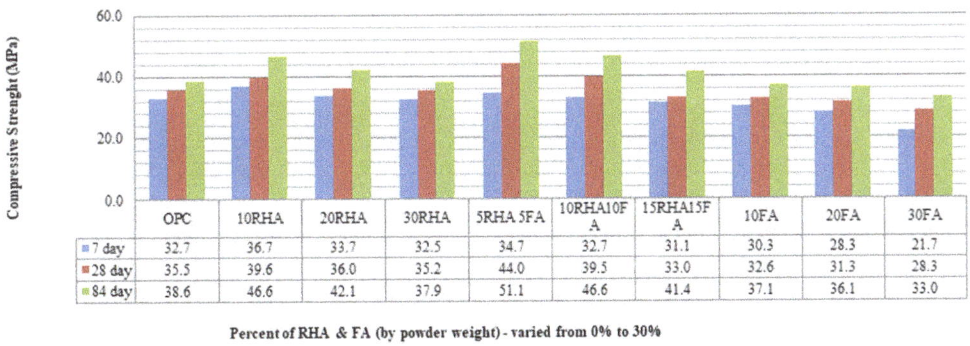

	OPC	10RHA	20RHA	30RHA	5RHA 5FA	10RHA10FA	15RHA15FA	10FA	20FA	30FA
7 day	32.7	36.7	33.7	32.5	34.7	32.7	31.1	30.3	28.3	21.7
28 day	35.5	39.6	36.0	35.2	44.0	39.5	33.0	32.6	31.3	28.3
84 day	38.6	46.6	42.1	37.9	51.1	46.6	41.4	37.1	36.1	33.0

as slower pozzolanic reaction of the FA with $Ca(OH)_2$ during the hydration process. Moreover, chemical reaction and strength gain rate for concrete containing FA was comparably slower at early ages of curing (Islam and Islam, 2010). The mortar grout mixture containing 30% of FA content has the lowest strength at all curing ages, as compared to all mixtures. Due to the larger amount of cement replaced, the pozzolanic reaction of FA may require longer curing time to gain strength.

Durability Test on Harden State Mortar Grout

Water Absorption Test

Figure 2 shows the results for immersed water absorption test. The addition of RHA increased the water absorption of the concrete specimens with its increasing percentage as cement replacement. The percentage of water absorption increases with an increment of RHA content; perhaps due to the fineness of RHA particles which are finer than OPC and due to its hygroscopic nature (Ganesan et al., 2008). It can also be seen that the addition of FA reduced the water absorption of the concrete specimen. Moreover, absorption was found to decrease with increasing percentage of FA in the mixtures. The water absorption for inclusion of only FA in concrete is lower than control mix at all curing age and decreases with the increasing amount of FA.

The decrease in water absorption indicates higher mortar durability. Therefore, it showed that, FA aids the durability up to 30% cement replacement in the mortar. The absorption continues to decrease in all mixtures with

Figure 2. Immersed water absorption test results

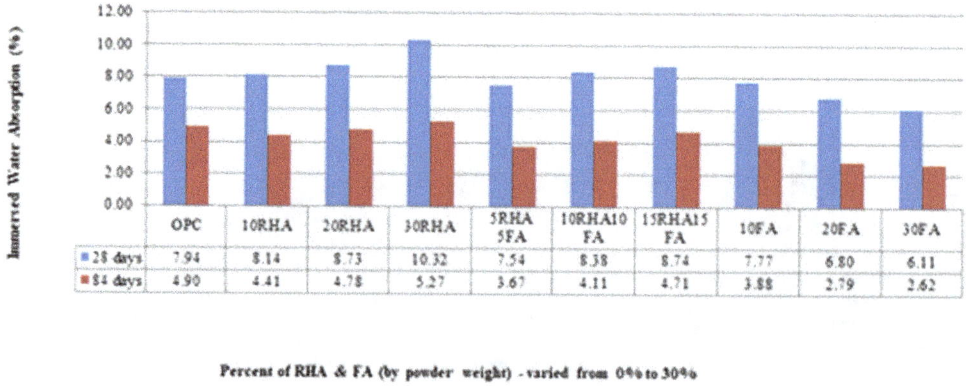

	OPC	10RHA	20RHA	30RHA	5RHA 5FA	10RHA10 FA	15RHA15 FA	10FA	20FA	30FA
28 days	7.94	8.14	8.73	10.32	7.54	8.38	8.74	7.77	6.80	6.11
84 days	4.90	4.41	4.78	5.27	3.67	4.11	4.71	3.88	2.79	2.62

curing periods from 28 to 84 days. This shows that the durability of mortar actually improved with time due to more pozzolanic reaction. Pozzolanic reaction will produce concrete with finer pore structure and hence increasing the durability of concrete.

Apparent Volume of Permeable Voids (AVPV) Test

Figure 3 shows the AVPV test result. It can be observed that mixtures containing only RHA show higher AVPV than control mix. In addition, increasing the RHA content, increased the AVPV values. According to Parande et al. (2011), increased replacement of RHA content causes the effective porosity to decrease. They attribute this mainly to the uncompleted formation of CSH gel formation in pozzolanic materials. RHA is a type of materials which also shows hygroscopic properties. RHA based mortars required more chemical admixtures, as the micro particles absorbed more water. Thus, the produced mortar grout was more permeable and less durable.

It appears that the addition of FA reduced the AVPV of concrete specimens and the absorption was found to decrease with increasing percentage of FA in the mixtures. The mixtures containing FA had lesser AVPV values than RHA mortars. This shows that FA is slightly more effective in modifying pore and reducing the permeability of mortar grout. Besides this, decreasing of AVPV values was observed with curing age and could be due to the slow hydration process in FA based grouts.

Figure 3. AVPV test results

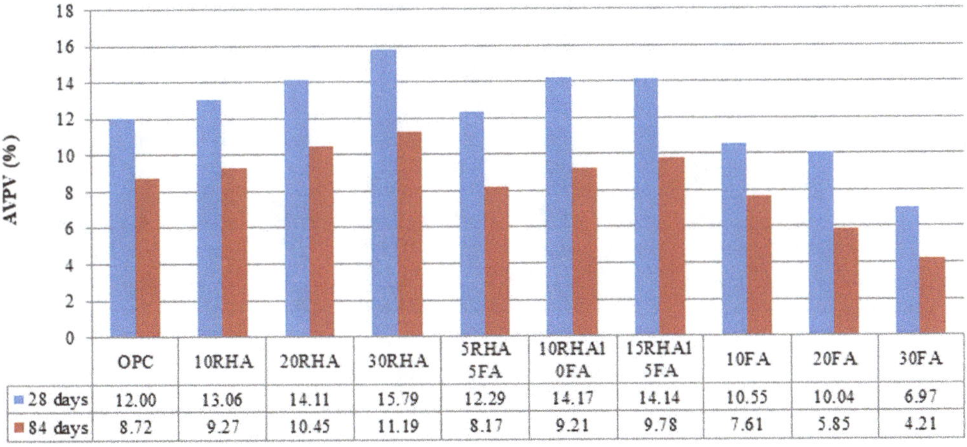

	OPC	10RHA	20RHA	30RHA	5RHA 5FA	10RHA10FA	15RHA15FA	10FA	20FA	30FA
28 days	12.00	13.06	14.11	15.79	12.29	14.17	14.14	10.55	10.04	6.97
84 days	8.72	9.27	10.45	11.19	8.17	9.21	9.78	7.61	5.85	4.21

Percent of RHA & FA (by powder weight) - varied from 0% to 30%

Sorptivity Test

Both Figures 4 and 5 show the initial and secondary rate of absorption decrease with increase in either RHA or FA as partial cement replacement in the mortar grouts. Similar results were reported by Ganesan et al. (2008), who mentioned that the sorptivity decreased continuously with increasing contents of RHA. Moreover, addition of FA in mortar grouts showed a lower rate of absorption as compared to that of control mix and mixtures containing RHA only.

The capillary pores of mixtures with FA concrete were filled with the aid of free lime (CaO) during the hydration process. Hence, the rate of absorption of mixture with FA was lower than the control mix due to the reduction of capillary pores. Voids that are filled in a packed system would reduce the sorptivity due to improvement of arrangements of particles in the system (Yahia et al., 2005).

It was also observed that sorptivity of the mortar grout decreases from the age of 28 days to the age of 84 days. Similar results were reported by Imbin et al. (2013). With prolonged curing time, more hydration process will take place and thus leads to a reduction in pore space and thus decrease the water uptake of the mortar grout.

Figure 4. Initial rate of absorption test results

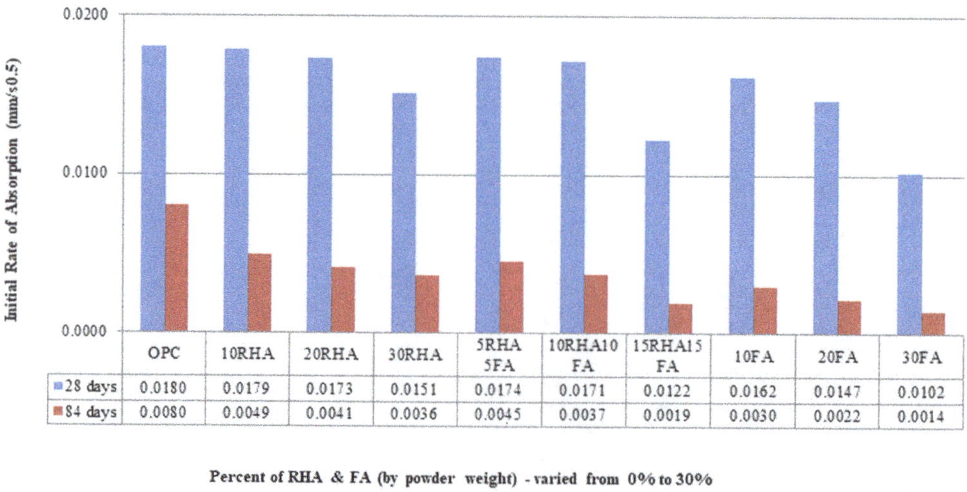

	OPC	10RHA	20RHA	30RHA	5RHA 5FA	10RHA10 FA	15RHA15 FA	10FA	20FA	30FA
28 days	0.0180	0.0179	0.0173	0.0151	0.0174	0.0171	0.0122	0.0162	0.0147	0.0102
84 days	0.0080	0.0049	0.0041	0.0036	0.0045	0.0037	0.0019	0.0030	0.0022	0.0014

Figure 5. Initial rate of absorption test results of mortar grout with various percentages replacement of RHA and FA in different curing time

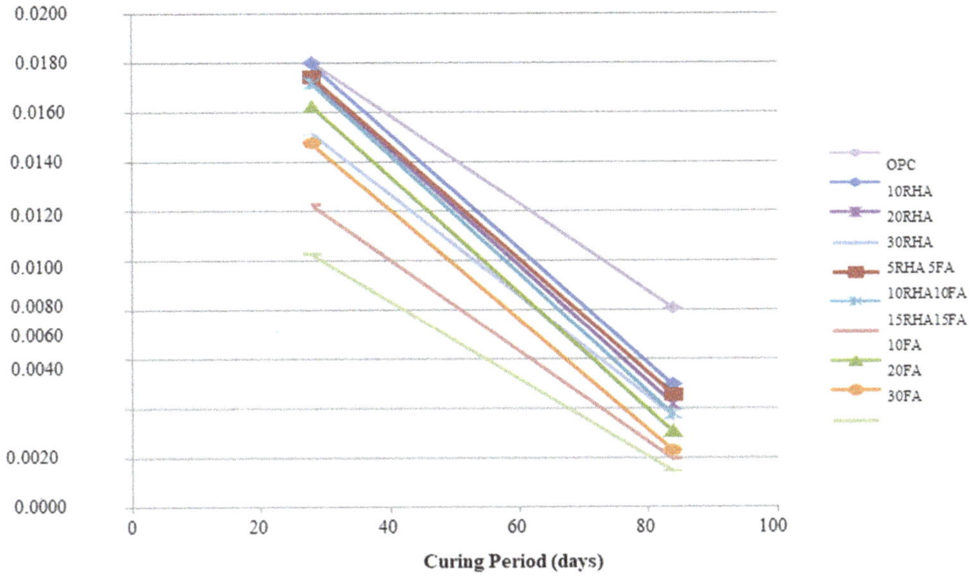

Rapid Chloride Ion Penetration (RCPT) Test

Results in Figure 6 show that addition of RHA in mortar grout enhances the resistance to chloride penetration as compared to control mix and mixtures with FA. The addition of RHA and FA in concrete as partial cement replacement material increases nucleation sites for precipitation of hydration products which reduces the $Ca(OH)_2$ and improves the permeability of concrete (Chindaprasirt et al., 2008). These factors help in developing resistance improvement of chloride penetration in concrete, which the RHA being the best, followed by FA.

It can be observed that total Coulombs charge passing through mixtures containing RHA decreased with the increase in RHA content up to 30%. The lower un-burnt carbon content found in RHA would have helped in greatly reducing the total Coulombs charge passed since the charge passing through concrete relies on the electrical conductance where unburnt carbon might act as a medium for conductivity (Ganesan et al., 2008). Besides, the inclusion of FA in mortar grout also reduced the total Coulombs charge passed with its increasing content in the mixtures.

Figure 6. RCPT test results

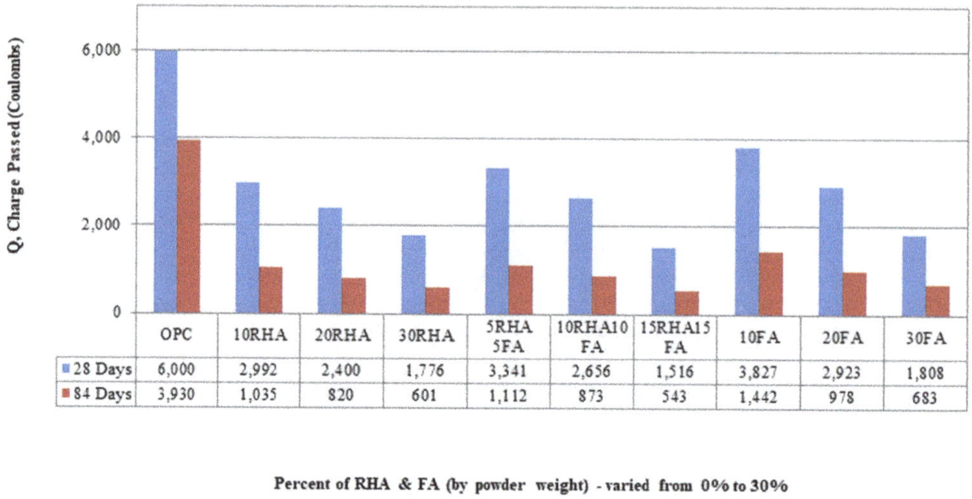

Resistance of chloride ion penetration is expressed as chloride resistivity in mortar grout. The inclusion of FA in mortar grout has high electrical resistances and thus lower permeability in mortar grouts (Boga and Topcu, 2012). With the continuous water curing of up to 84 days, the charge passed through all mortar grouts can be significantly reduced.

Chloride resistivity of concrete is also related to pore structure of the cement paste matrix. A longer curing time causes more hydration reaction to take place in developing the pore structure of mortar grouts. In other words, it leads to an increase in chloride resistivity or decreases the diffusion of chloride ions in mortar grouts.

CONCLUSION

For the mechanical properties of mortar grout, the inclusion of blended RHA and FA significantly improves its compressive strength. The maximum compressive strength was obtained by a combination of 5% of RHA and 5% of FA in mortar grout at 84 days. Further increasing the RHA and FA content, reduces the compressive strength of mortar grout. Moreover, the use of mortar grout containing RHA and FA improve its durability in the long term. The durability of mortar grout increases along with a longer curing time as more hydration will occur to refine the concrete pore structures. Hence, RHA and FA which are locally available in Malaysia can partially replace cement in the production of mortar grout.

ACKNOWLEDGMENT

We, the editors greatly appreciate the exceptional contribution of Dr. Muhammad Ekhlasur Rahman for carrying out this research and writing of the chapter. Indeed, members of his team, Dr. Brabha H. Nagaratnam, Dr. Han Seung Lee, and Mr. Shafiq Ishak deserve to be equally acknowledged for their co-authorship and immense collaboration in conducting/writing of the research.

REFERENCES

Al-Khalaf, M. N., & Yousift, H. A. (1984). Use of rice husk ash in concrete. *The International Journal of Cement Composites and Lightweight Concrete*, *6*(4), 241–248. doi:10.1016/0262-5075(84)90019-8

ASTM International-2012. *Standard Specification for Coal Fly Ash and Raw or Calcined Natural Pozzolan for Use in Concrete.* (ASTM C618 - 12).

ASTM International-2012. *Standard Test Method for Electrical Indication of Concrete's Ability to Resist Chloride Ion Penetration.* (ASTM C1202 – 12).

ASTM International-2012. *Standard Test Method for Measurement of Rate of Absorption of Water by Hydraulic-Cement Concretes.* (ASTM C1585 - 11).

ASTM International-2013. *Standard Test Method for Slump of Hydraulic-Cement Concrete.* (ASTM C143/C143M - 12).

Boğa, A. R., & Topçu, İ. B. (2012). Influence of fly ash on corrosion resistance and chloride ion permeability of concrete. *Construction & Building Materials*, *31*(0), 258–264. doi:10.1016/j.conbuildmat.2011.12.106

Chindaprasirt, P., Rukzon, S., & Sirivivatnanon, V. (2008). Resistance to chloride penetration of blended portland cement mortar containing palm oil fuel ash, rice husk ash and fly ash. *Construction & Building Materials*, *22*(5), 932–938. doi:10.1016/j.conbuildmat.2006.12.001

Gambhir, M. L. (2004). *Concrete Technology* (3rd ed.). Civil Engineering Series, Tata McGrow-Hill.

Ganesan, K., Rajagopal, K., & Thangavel, K. (2008). Rice husk ash blended cement: Assessment of optimal level of replacement for strength and permeability properties of concrete. *Construction & Building Materials, 22*(8), 1675–1683. doi:10.1016/j.conbuildmat.2007.06.011

Givi, A. N., Abdul Rashid, S., & Aziz, F. N., & Salleh, M. A. M. (2010). Contribution of rice husk ash to the properties of mortar and concrete: A review. *The Journal of American Science*, *6*(3), 157–165.

Habeeb, G. A., & Mahmud, H. (2010). Study on properties of rice husk ash and its use as cement replacement material. *Materials Research, 13*(2), 185–190. doi:10.1590/S1516-14392010000200011

Horrigmoe, G. (2000). Future needs in concrete repair technology. In O. E. Gjorv & K. Sakai (Eds.), *In proceedings, Concrete Technology for a Sustainable Development in the 21st Century* (pp. 332–340). London: E & FN Spon.

Imbin, S., Dullah, S., Asrah, H., Kumar, P. S., Rahman, M. E., & Mannan, M. A. (2013). Performance of mortar grout under aggressive chloride environment in sabah. *World Academy of Science, Engineering and Technology*, *73*, 1171–1175. Retrieved from http://www.waset.org/journals/waset/v73/v73-212.pdf

Islam, M. D. M., & Saiful Islam, M. (2010). Strength behaviour of mortar using fly ash as partial replacement of cement. *Concrete Research Letters, 1*(3), 98-106.

Nagaratnam, B. H., Rahman, M. E., Mirasa, A. K., Mannan, M. A., & Lame, S. O. (2016). Workability and heat of hydration of self-compacting concrete incorporating agro-industrial waste. *Journal of Cleaner Production, 112*, 882–894. doi:10.1016/j.jclepro.2015.05.112

Naik, T. R., Singh, S. S., & Hossain, M. M. (1995). Properties of high performance concrete systems incorporating large amounts of high-lime fly ash. *Construction & Building Materials*, *9*(4), 195–204. doi:10.1016/0950-0618(95)00009-5

Nguyen, V. T., Guang, Y., Klass, V. B., & Oguzhan, C. (2011). Hydration and microstructure of ultra high performance concrete incorporating rice husk ash. *Cement and Concrete Research*, *41*(11), 1104–1111. doi:10.1016/j.cemconres.2011.06.009

Papadakis, V. G., & Tsimas, S. (2002). Supplementary cementing materials in concrete part i: Efficiency and design. *Cement and Concrete Research*, *32*(10), 1525–1532. doi:10.1016/S0008-8846(02)00827-X

Parande, A. K., Stalin, K., Thangarajan, R. K., & Karthikeyan, M. S. (2011). Utilization of agroresidual waste in effective blending in portlant cement. *ISRN Civil Engineering*, *2011*, 1–12. doi:10.5402/2011/701862

Rahman, M. E., Muntohar, A. S., Pakrashi, V., Nagaratnam, B. H., & Sujan, D. (2014). Self compacting concrete from uncontrolled burning of rice husk and blended fine aggregate. *Materials & Design*, *55*, 410–415. doi:10.1016/j.matdes.2013.10.007

Reddy, D. V., & Alvarez, M. (2006). Marine durability characteristics of rice husk ash-modified reinforced concrete. *Proceedings of Fourth LACCEI International Latin American and Caribbean Conference for Engineering and Technology (LACCET'2006).*

Sanal, I. (2017). A review on reduced environmental impacts of alternative green concrete productions. *International Journal of Public and Private Perspectives on Healthcare, Culture, and the Environment*, *1*(2), 55–68. doi:10.4018/IJPPPHCE.2017070104

Sanal, I. (2018). Discussion on the effectiveness of cement replacement for carbon dioxide (CO_2) emission reduction in concrete. *Science and Technology*, *8*(2), 1748. doi:10.1002/ghg

Sanal, I. (2018). Significance of concrete production in terms of carbon dioxide emissions: social and environmental impacts. *Journal of Polytechnic*, *21*(1). Doi:10.2339/politeknik.389590

Sathawane, S. H., Vairagade, V. S., & Kene, K. S. (2013). Combine effect of rice husk ash and fly ash on concrete by 30% cement replacement. *Procedia Engineering*, *51*(0), 35–44. doi:10.1016/j.proeng.2013.01.009

Somayaji, S. (2001). *Civil Engineering Materials* (2nd ed.). Prentice Hall, Inc.

Standards Australia. (1999a). Methods of Testing Concrete: Method 9: Determination of the Compressive Strength of Concrete Specimens. SAIGlobal (AS 1012.9-1999)

Standards Australia. (1999b). Methods of Testing Concrete: Method 21: Determination of Water Absorption and Apparent Volume of Permeable Voids in Hardened Concrete. SAIGlobal (AS 1012.21-1999)

Tateda, M., Sekifuji, R., Yasui, M., & Yamazaki, A. (2016). Case study: Technical considerations to optimize rice husk burning in a boiler to retain a high solubility of the silica in rice husk ash. *Journal of Scientific Research & Reports*, *11*(4), 1–11. doi:10.9734/JSRR/2016/27902

Uduweriya, R. B. Y. B., & De Silva, S. (2010). *Effect of mechanical properties of concrete containing rice husk ash*. University of Ruhuna. Retrieved from http://www.ruh.ac.lk/research/academic_sessions/2011_mergepdf/bf4.pdf

Yahia, A., Tanimura, M., & Shimoyama, Y. (2005). Rheological properties of highly flowable mortar containing limestone filler-effect of powder content and w/c ratio. *Cement and Concrete Research*, *35*(3), 532–539. doi:10.1016/j.cemconres.2004.05.008

Zain, M. F. M., Islam, N., Mahmud, F., & Jamil, M. (2010). Production of rice husk ash for use in concrete as supplementary cementitious material. *Construction & Building Materials*, *25*(2), 798–805. doi:10.1016/j.conbuildmat.2010.07.003

Chapter 4
Effect of Metakaolin on Geopolymer Industry

ABSTRACT

This chapter discusses the effects of metakaolin (MK) on geopolymer mortar and concrete industries. The research topics of MK-based geopolymer cover reaction mechanisms and kinetics. This chapter aims at augmenting knowledge about enhancing mechanical properties of geopolymer mortars/concrete using MK. Specifically, this chapter presents literature studies as well as current experimental studies which delineate the effect of MK on fresh and hardened-state properties of geopolymer mortars (GPMs). Properties and characteristics of metakaolin are explained followed by properties of fresh MK mortars. Properties of hardened MK concrete and durability aspects of MK mortars are explained. Applications of MK-based geopolymers and metakaolin-based geopolymers as repair materials are also included in this chapter. The results of using MK-based GPMs revealed improved workability, enhanced setting time, increased density, higher compressive strength, flexural strength, and resistance against acid attack than conventional ordinary portland cement mortar/concrete.

INTRODUCTION

In late 1970s, Professor Joseph Davidovits carried out a research on fireproof polymers and introduced the concept of geopolymer, which was coined to describe a family of alkali activated alumina-silicate binders (Davidovits, 2008). The formation of geopolymer was based on the reaction between the two parts of materials: alkali activator and reactive alumina-silicate precursor.

DOI: 10.4018/978-1-5225-8325-7.ch004

In this study, the precursor was mainly metakaolin (MK) and is at a very early stage of the research development. The geopolymer based on alkali activation of MK has gained worldwide interests in the past several years. It was not only because of its excellent thermal stability, which is much better compared with conventional polymer material, but also due to its comparable mechanical properties to cements, which is being considered as a green alterative to Portland cement (Duxson et al., 2007). In recent years, the rapid growth in research and development of geopolymer has shifted from the early interests of thermal resistant applications towards construction and building materials (Pacheco-Torgal et al., 2008). Because of this, the raw materials used for large volume geopolymer manufacture has been significantly broadened, including heated low quality clays and a variety of silica and aluminium-bearing (Si- and Al) waste materials and by-products sourced from different industries (Zhang et al., 2016).

Despite the fact that fly ash and slag became the two major waste materials used in today's limited commercial geopolymer products, MK is probably still the most promising feedstock materials for geopolymer in the future. This is because MK has more consistent chemical compositions than fly ash and slag, and is expected to result in more consistent and predictable products. In fact, fly ash and slag are becoming less available in many countries because of their effective usage in the manufacture of blending cements and concrete (Malhotra and Mehta, 1996). Therefore, from a long-term point of view, the use of MK as raw material is becoming more attractive and realistic. MK is a thermally treated product from kaolin, which is one of the naturally occurring abundant minerals in the earth's crust (Zhou and Keeling, 2013). Kaolin has been historically used in the production of Portland cement. To produce one tonne of clinker, the most important ingredient of cement, it requires about 0.3 tonnes of clay to mix with 1.5 tonnes of limestone and other iron-bearing minerals and calcination at 1450 °C (Zhang et al., 2014). Kaolin can also be used in another form as the supplementary cementing material in concrete mixing. This utilization requires a thermal treatment process, usually at temperatures ranging between 500 °C and 800 °C, to convert kaolin into metakaolin (MK). MK is a pozzolan, which in itself possesses little or no cementing property but will react chemically with calcium hydroxide ($Ca(OH)_2$) to form compounds in the presence of water to possess cementing properties. Theoretically, the replacement, usually 5–20% by mass of cement, contributes to a slight reduction in CO_2 emission due to the less intensive thermal and grinding treatments required in obtaining it compared with cement clinker. In practice, however, MK is not commonly used in most construction cases due to its relatively large specific surface area, which may demand high water/binder ratio to achieve satisfying workability. This means that such a family of abundant resource is not utilized in large volume in concrete, the largest man-made material used in the world. In comparison, geopolymer can be made with 100% MK, and this type of binder can reduce by up to 70–80% of CO_2 emissions.

Metakaolin has been the preferred alumino-silicate material among researchers (Barbosa and MacKenzie, 2003; Duxson et al., 2007) due to its high rate of dissolution in the reactant solution, ability to manufacture with same homogenous properties, even though it needs extensive energy to produce. It was recorded that about 300,000 tonnes of locally produced calcined clay (metakaolin) were used in Amazon basin in 1960s (Sabir et al., 2001). With the worldwide interests of geopolymer, the number of scientific publications has increased exponentially in recently years. The research topics of MK-based geopolymer cover from the reaction mechanisms to kinetics.

The aim of this book chapter was to increase the knowledge on the enhancement of mechanical properties of geopolymer mortars/concrete using MK. Specifically, this chapter presents the studies from the literature as well as our experimental studies on the effect of MK on fresh- and hardened-state properties of geopolymer mortars (GPMs).

Properties and Characteristics of Metakaolin

Physical Properties

Calcination of clay mineral (kaolin) produces metakaolin at moderately-high temperatures (650-800°C). Metakaolin particles, off-white in color, are extremely small with an average particle size of 3-10 μm (Figure 1). Some physical properties of metakaolin are given in Table 1. The field emission

Table 1. Physical properties of metakaolin

Ref.	Specific Gravity	Fineness m^2/kg	Partial Size d50 μm	Colour
(Tafraoui et al., 2009)	2.50	15000	12.0	-
(Al-Akhras 2006)	2.50	12000	1.0	white
(Poon et al., 2001)	2.62	12680	-	-

Figure 1. Metakolin

scanning electron microscopy (SEM) images, taken for the metakaolin samples, revealed irregular platy and angular shaped particles. These are closely packed in lumps with certain areas disorderly arranged as depicted in Figure 2. The tiny lumps observable in their physical form which could be noticed in the micrograph as packs of layers are of metakaolin sheets.

Figure 2. SEM of metakolin sample

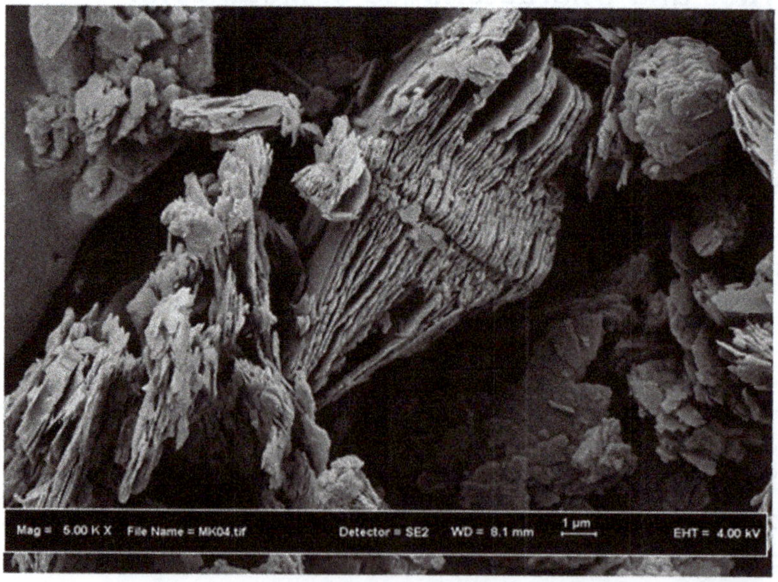

Chemical Compositions

Major constituents of metakaolin are silica oxide (SiO_2) and alumina oxide (Al_2O_3). Other components include ferric oxide, calcium oxide, magnesium oxide, potassium oxide, etc. The typical chemical composition of metakaolin is given in Table 2. Results revealed that the major constituents of metakaolin are silica oxide (SiO_2 between 50 and 75%) and aluminum oxide (Al_2O_3) ranging between 27 and 42%. Metakaolin should meet the requirements of ASTM C618 (ASTM 1999), "Standard Specification for Coal Fly Ash and Raw or Calcined Natural Pozzolan for use as a mineral admixture in Concrete, Class N", with the following modifications as given in Table 3.

The metakaolin used was the product of heat-treated kaolin at 650°C for 4 h. The chemical composition and specific weight data of the original kaolin and metakaolin are given in Table 4 (as provided in different research papers).

Table 2. Chemical compositions of MK by using XRF test

Ref.	SiO_2	Al_2O_3	CaO	MgO	Fe_2O_3	Na_2O	K_2O	TiO_2	Others	LOI	Si+Al	Si:Al
(Tchakouté et al., 2016)	54.50	27.40	0.10	0.09	2.55	0.09	0.28	2.21	1.59	11.19	81.90	1.98
(Alanazi et al., 2016)	55.01	40.94	0.14	0.34	0.55	0.09	0.60	0.55	0.24	1.54	95.95	1.34
(Reig Cerdá et al., 2016)	58.39	35.47	0.01	0.30	2.71	-	1.44	1.51	0.07	0.10	93.86	1.66
(Cyr and Pouhet 2016)	68.10	24.10	0.91	0.22	3.73	0.08	0.35	1.14		1.37	92.20	2.83
(Duan et al., 2015)	53.32	42.09	0.09	0.21	2.33	0.49	0.64	0.63	0.12	0.08	95.41	1.26
(Gordon et al., 2014)	52.8	39.2	0.10	0.20	0.80	-	2.7	-	3.0	1.2	92.0	1.33
(Arellano-Aguilar et al., 2014)	75.41	22.91	-	-	0.64	-	0.52	0.52	-	-	98.32	3.40
(Ismail et al., 2013)	54.7	39.9	-	0.70	1.43	-	2.58	-	5.4	-	94.6	1.37
(Živica et al., 2011)	53.68	42.01	0.21	0.10	1.70	-	-	-	2.30		95.69	1.28
(Rovnaník 2010)	55.01	40.94	0.14	0.34	0.55	0.09	0.60	0.55	0.24	1.54	95.95	1.34
(Tafraoui et al., 2009)	58.10	35.14	1.15	0.20	1.21	0.07	1.05	-	1.23	1.85	93.24	1.65
(Wild and Khatib 1997)	52.10	41.0	0.07	0.19	4.32	0.26	0.63	0.81	0.02	0.60	93.10	1.27
(Ambroise et al., 1994)	51.52	40.18	2.0	0.12	1.23	0.08	0.53	2.27	0.06	2.01	91.70	1.27

Table 3. Requirements of metakaolin (ASTM C 618)

Modified Specification Requirements	
Item	**Limit**
Silicon dioxide (SiO_2) plus aluminum oxide (Al_2O_3) plus iron oxide (Fe_2O_3)	Min 85%
Available alkalis	Max 1.0%
Loss on ignition	Max 3.0%
Fineness: amount retained when wet-sieved on 45-µm sieve	Max 1.0%
Strength activity index at 7 days (% of control)	85
Increase of drying shrinkage of mortar bars at 28 days	Max 0.03%

Table 4. Chemical composition and properties of kaolin and metakaolin

Chemical Composition (%)		
Components	**Kaolin**	**Metakaolin**
Humidity	0.94	0.35
Loss on Ignition	12.20	1.95
SiO_2	48.06	53.68
Al_2O_3	36.76	42.0
Fe_2O_3	1.42	1.7
CaO	0.33	0.21
MgO	0.24	0.10
Specific weight ($kg.m^{-3}$)	2631	2582
Solution heat (Jg^{-1})	278.2	5563

(Živica et al., 2011)

With the aim to compare the reactivity of both materials also, their solution heat was estimated using the kinetics method of its development (Živica et al., 2011). The comparison of chemical composition showed an increase in the SiO_2, Al_2O_3 and Fe_2O_3 contents of the metakaolin. This was the direct consequence of the decrease in loss on ignition occurred due to heat-treatment of kaolin. Further interesting effects observed were a significant (approximately 20 times) increased solution heat and, on the contrary, a decreased specific weight of the metakaolin. The increased solution heat indicated the increased reactivity of metakaolin opposite to kaolin. The decrease in specific weight indicated the decreased compactness of metakaolin. This effect was confirmed by the porosimetry results showing 31.8% increase of pore volume and 20.2% increase of total porosity. Furthermore, 40.7% decrease in macro-pores' contents and 68.9% increase in specific surface area indicated the metakaolin as a material significantly finer than the starting kaolin.

Figure 3 shows the XRD pattern of MK. MK comprised of mainly amorphous phase as exhibited by the broad hump around 18–62° with crystalline phases of mullite ($Al_6Si_2O_{13}$), quartz (Si), (Mg_2Si), ($CaSiO_2$) and (Al_4Ca).

All the mentioned effects were undoubtedly the consequence of the heat-treatment process, producing the amorphous or near-amorphous SiO_2–Al_2O_3– Fe_2O_3 products. Its presence in the metakaolin is confirmed from the DTA-curves by a very broad endotherm between 200 and 950°C, instead of the sharp endotherm peak at 511°C visible on the DTA-curve of the kaolin.

Hydration Reaction

In Portland cement concrete, MK reacts at normal temperatures with calcium hydroxide in cement paste to form mainly calcium silicate hydrates (C–S–H), C_2ASH_8 (gehlenite hydrate), and C_4AH_{13} (tetra calcium aluminate hydrate). The formation of secondary C–S–H by this reaction with kaolin reduces the total porosity and refines the pore structure thus, improving the strength and impermeability of the cementitious matrix.

In geopolymer concrete, the reactions are different. The polymerization process involves a substantially fast chemical reaction under alkaline condition

Figure 3. XRD of metaokin sample

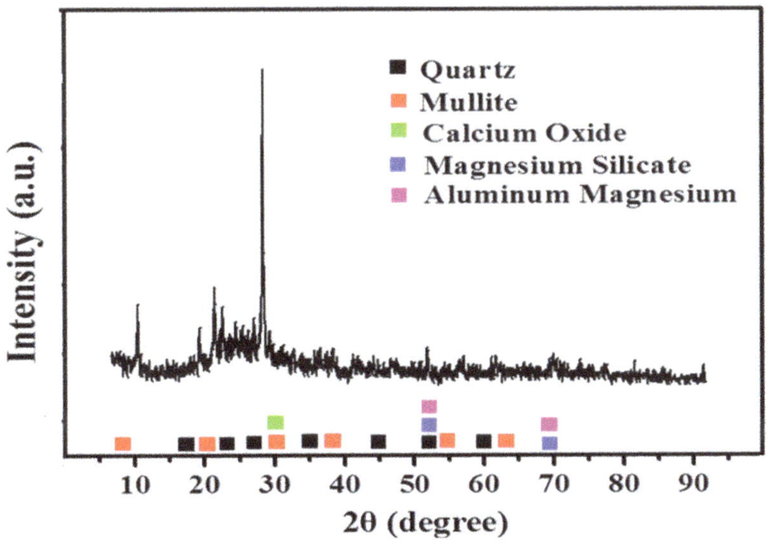

on Si -Al minerals, that results in 3D polymeric chain and ring structure consisting of Si-O-Al-O bonds. The main concept behind this geopolymer is the polymerization of the Si-O-Al-O bond which develops when Al-Si source materials like fly ash or rice husk ash is mixed with alkaline activating solution (NaOH or KOH solution with Na_2SiO_3 or K_2SiO_3). The main product of geopolymerization process is sodium aluminosilicate hydrate (NASH). The geopolymer can be in the form of -Si-O-Al-O- or –Si-O-Al-O-Si-O- or –Si-O-Al-O-SiO-Si-O-. The chemical reaction behind the formation of geopolymer areas is as follows in Figure 4.

From the above equation in Figure 4, it is clear that water does not play any role in the polymerization process because it was expelled from the chemical reaction.

Properties of Fresh MK Mortars

Workability and Setting Time

Two different tests, such as, workability and setting time were reported to evaluate the fresh properties of geopolymer mortars (GPMs). In our experimental study, it was found that the metakaolin contributed to the improvement in workability of geopolymer mortar. The workability of the

Figure 4. Impact of MK substituting GBFS on workability of GPMs

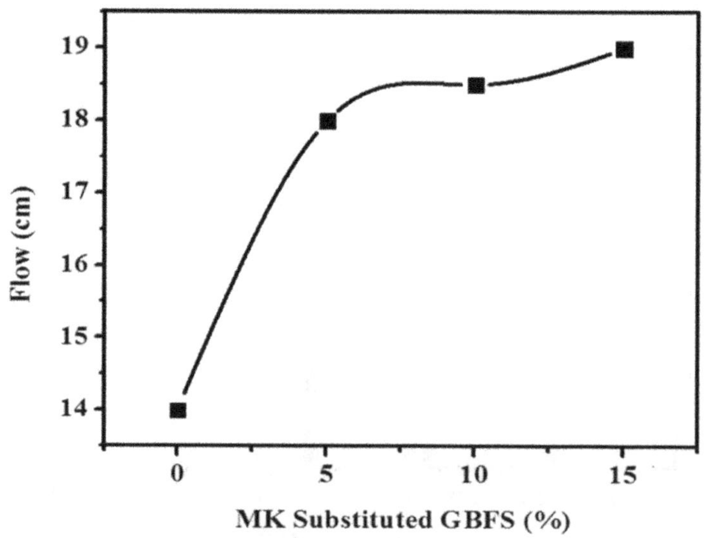

mortar increased with the increase in MK percentage as replacement of GBFS as illustrated in Figure 5. This could be attributed to the differences in the physical properties and chemical reactions of the mixtures. Furthermore, with the reduction of GBFS content, the number of angular particles reduced which helped to improve the workability of the mortar mixture. On top, admixing MK and GBFS produced slow setting and enhanced workability as shown in Figure 6. This observation was also supported by the findings of Majidi et al., (Al-Majidi et al., 2016). It was also observed that a decrease in calcium contents and increase in Si and Al contents led to enhance the initial and final setting times as reported elsewhere (Lee et al., 2016; Phoo-ngernkham et al., 2015; Sofi et al., 2007). Furthermore, an increase in the MK content contributed to increase the SiO_2 and Al_2O_3, which led to improve the setting time (Chindaprasirt et al., 2012). The rate of setting time increased significantly as indicated by the substantial differences in the initial setting time. The difference between initial and final setting times also increased with the reduction of GBFS content in the mortar. It also supported the fact that higher the GBFS content in the mortar, the quicker is the rate of setting (Kumar et al., 2010; Sugama et al., 2005). Thus, it could be established that MK, as part of the binary blended binder, is very effective to decelerate the setting time of GPMs under ambient condition.

Figure 5. Impact of MK substituting GPES on setting times of GPMs

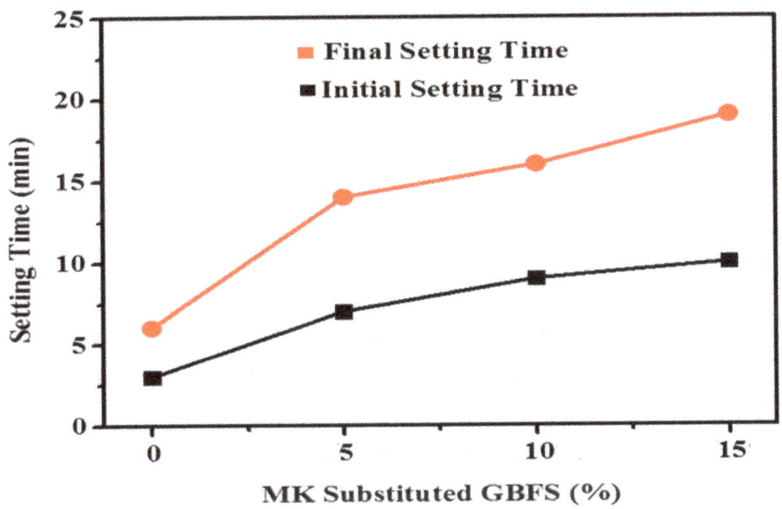

Figure 6. Effect of MK substituted GPFS on GPMs density

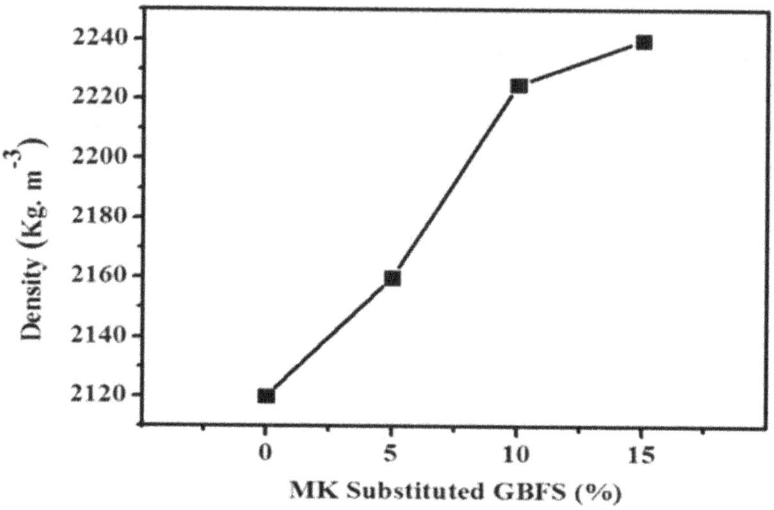

Effect of Metakaolin on Geopolymer Density

Figure 7 illustrates the effect of MK substituted GBFS on the density of GPMs. The density of GMPs was found to increase with increasing percentage of MK as replacement of GBFS. The particle size and specific gravity of MK was found to influence the GPMs density. Also the increased contents of Al_2O_3 and SiO_2 led to produce sodium aluminosilicate hydrate (NASH) gel besides the calcium silicate hydrate (CSH), This improved the microstructure of GPMs as reported in previous research studies (Huseien et al., 2016a; Huseien et al., 2016b).

Properties of Hardened MK Concrete

Effect of Metakaolin Content on Compressive Strength

Görhan et al. (2016) studied the substitution of fly ash by metakaolin ranging from 10% to 40% to produce geopolymer mortar. It was reported that the increase in metakaolin content improved the compressive strength of GPM from 18 MPa to 26 MPa. The hardened properties of metakaolin based geopolymer replaced by sewage sludge ash (10 and 20%) was investigated by Reig et al. (2016). The results indicated that the reduced MK content affected negatively

Figure 7. Effect of MK substituted GBFS on GMPs density

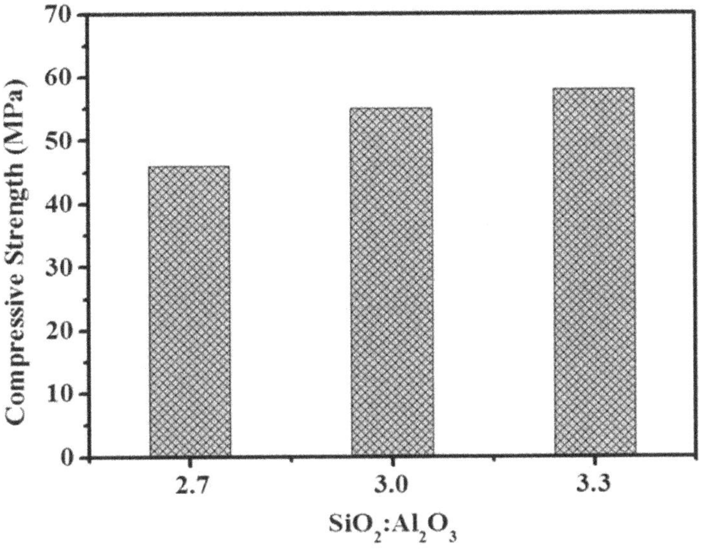

in strength properties of GPMs. The compressive strength dropped from 37 MPa to 26 MPa with 100 and 80% of metakaolin content, respectively.

Alanazi et al. (2016) reported the influence of metakaolin on bond strength between GPM and cement concrete whereas effect of Si: Al in metakaolin composition was evaluated by Arellano-Aguilar et al. (2014). They reported that the increase in silicate content from 2.7 to 3.3 improved the compressive strength rate. The geopolymer prepared based metakaolin displayed a good bond strength with cement concrete using splitting and slant shear tests which reached 3.63 MPa and 16.32 MPa, respectively. The "h" value is 26% as shown in Figure 8.

Chen et al. (2011) also published the reduction in compressive strengths of geopolymers with increasing replacement of metakaolin by calcined reservoir sludge (CRS) particles. As compared to metakaolin, the solubility and concentration of geopolymer precursors of less reactive CRS particles in an alkaline activating solution are lower as displayed in Figure 9.

Effect of Curing Regime on GPMs Microstructure

In geopolymer industry, curing regime plays an important role in the geopolymerization and building of microstructure of hardened samples. Curing scheme includes the factors, such as temperature, duration and humidity. MK-

Figure 8. Effect of SiO2Al2O2 content on compressive strength of MK based on GMPs
Source: Arellano-Aguilar et al, 2014)

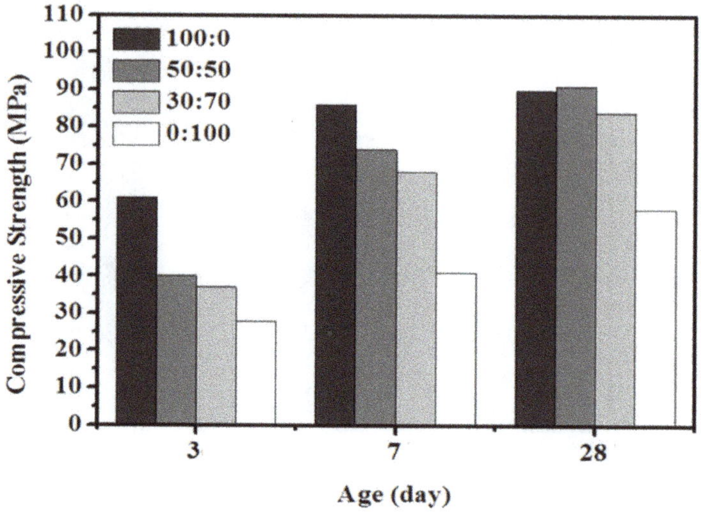

Figure 9. Comparison of the compressive strength of MK replaced by CRS based on GPMs
Source: Chen et al, 2011

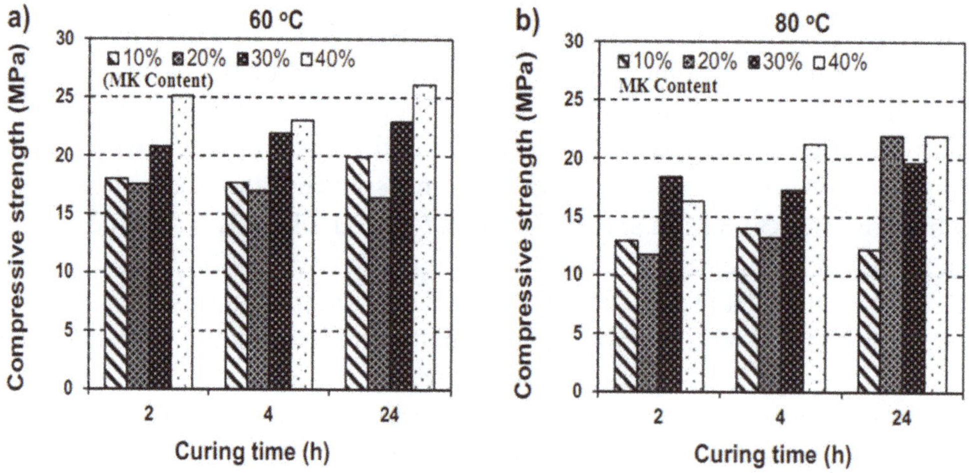

based geopolymers are usually prepared at elevated temperatures. The effect of curing temperature is discussed in this section. The GPM samples were cured at elevated temperatures of 65°C which displayed a higher strength compare to the samples cured at ambient temperature (27°C). Compressive strength dropped from 37 to 26.4 MPa when curing temperature was reduced as mentioned above (Reig et al., 2016). Gorhan et al. (2016) reported the effect of two oven temperatures: 60 and 80°C on the compressive strength development. Experimental results indicated that the GPM samples cured at high temperatures (80°C) displayed lower strength compare to other samples cured at lower temperatures as shown in Figure 10. The effect of curing period was also investigated by Chen et al. (2014) and reported that the increase in curing time from 2 to 24 hours improved the strength properties of GPMs.

Durability Properties of MK Mortars

Several studies (Alanazi et al., 2016; Kong et al., 2007) showed that the metakaolin have positive effects on durability of GPMs, such as increase in resistance of geopolymers' concrete to elevated temperatures, enhance the resistance against sulphate and acid attacks, etc. Duan et al. (2015) reported that the metakaolin mixed with fly ash could reduce the water absorption and increase the resistance to high temperatures. Other authors (He et al., 2010; Pan et al., 2014; Ranjbar et al., 2014) explored some basic aspects of metakaolin based geopolymers activated using sodium alkali solution for high-temperature applications. They estimated that the structure was stable enough to resist high temperatures.

Figure 10. Effect of curing temperature and curing period on GPMs strength and properties
Source: Gorham et al, 2016

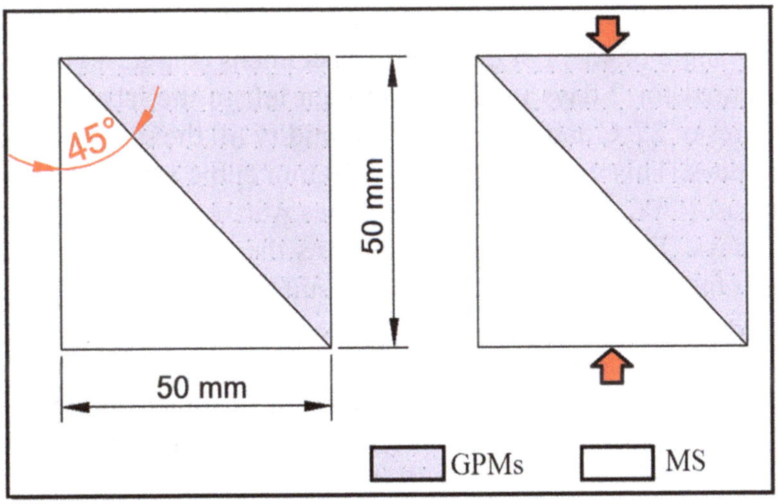

Jin et al. (2016) found in his investigation and his results indicated that metakaolin based geopolymer presented excellent stability in the acid and alkaline environments. Hence, there could be a potential for GPMs as building materials that could perform extremely well in aggressive environments with no secondary affects. Pateech et al. (2012) studied the acid, sulphate and freezing-thawing resistance of metakaolin based geopolymer and reported important advantages of using MK in geopolymer industry.

Applications of MK Based Geopolymers

Many potential applications have been proposed for the GPMs, such as, sustainable geopolymer concrete, fire-proof building materials and environmental materials, etc. This section summarises the very promising applications that have been extensively reported by researchers.

- Anti-corrosive cement and concrete products
- Thermal resistant and fire-proof materials
- Advanced inorganic composites and ceramics
- Waterproof, fast setting geopolymer repair materials
- Using MK as repair material for concrete surface especially exposed to aggressive environments, etc.

Metakaolin Based Geopolymers as Repair Materials

The effect of MK on GPMs bond strength was assessed as repair materials. ASTM C882 (ASTM 1991) and was utilized to evaluate the shear bond strength capacity between the Portland cement substrate (MS) and GPM with stiffer slant shear angle of 45°. For casting the specimens (Figure 11), the MS was cast and cured for 3 days in water and then left in the laboratory ambient temperatures of 27°C and 75% relative humidity till the samples reached the age of 28 days. This was then placed in 50 mm cubic molds and the second part was cast (OPC and GPM). These cubes were evaluated at ages of 1, 3, 7 and 28 days. The shear bond strength was the ratio of maximum load at failure and the bond area. The reported results of shear bond strength were the average of three samples.

The bond strength between OPC and GPM was determined using a slant shear bond test. Cubical slant shear specimens of dimension (50 mm × 50 mm × 50 mm) were prepared with interface line at 45° (Figure 11). They were

Figure 11. Preparation of GPMs samples for bound strength test

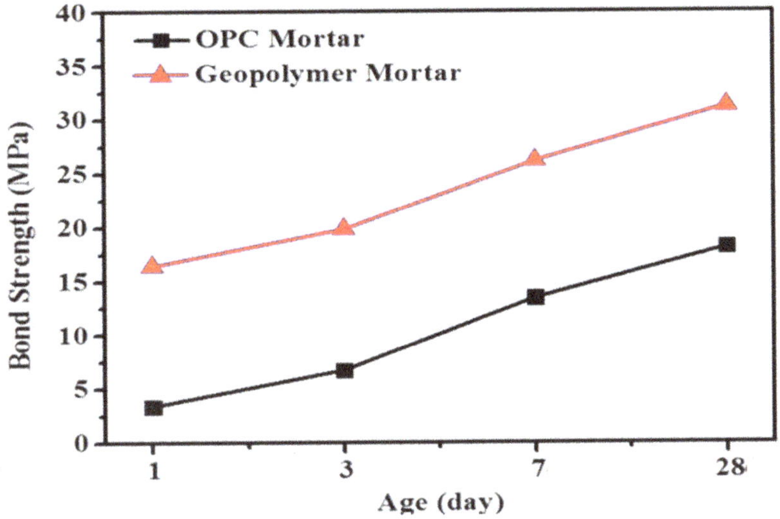

tested under compression using the standard procedure. The slant shear test is the most widely accepted test for the bonding of repair materials to damaged concrete substrates. The results of GPMs bond strength were compared to OPC mortar as shown in Figure 12. The bond strength of GPM prepared with 5% weight MK, 14 M NaOH solution and 0.40 of S: B displayed the highest bond strength of 16.5 MPa and 31.4 MPa at the early age of 1 day and at the later age of 28 days, respectively compared to OPC mortar. Figure 13 illustrates the typical bond failure of a slant shear sample, however, its

Figure 12. Sheer bound strength between OPC mortar and GPMs with interface line at 45 to the vertical

Figure 13. Typical failure model of the GMPs as repair materials

bonding surface was still intact. The cracks passed through the MS substrate and GPM interface and there was no significant gap between the two bonding surfaces. Similar trends were also reported elsewhere (Huseien et al., 2017; Huseien et al., 2016b; Phoo-ngernkham et al., 2015).

CONCLUSION

Based upon our and literature studies, the following conclusions could be presented:

- GPMs increase compressive and flexural strengths
- GPMs reduce permeability (including chloride permeability)
- GPMs has a potential to reduce efflorescence. This occurs when calcium ions are transported by water to the surface. There they combine with CO_2 of the atmosphere to make calcium carbonate, which precipitates on the surface as a white residue.
- GPMs increase resistance to chemical attack
- GPMs increase durability
- GPMs reduce effects of alkali-silica reactivity (ASR)
- GPMs enhance workability and finishing of concrete
- GPMs reduce shrinkage, due to "particle packing" making concrete denser
- GPMs attain high bond strength indicating their ability as alternative potential repair material.

ACKNOWLEDGMENT

We, the editors greatly appreciate the exceptional contribution of Dr. Ghasan Fahim Huseien for carrying out this research and writing of the chapter. Indeed, members of his team, Dr. Mohammad Ismail, Dr. Mohammad Ali Asaad, Dr. Kharunisa Muthusamy and Dr. Muhammad Ekhlasur Rahman deserve to be equally acknowledged for their co-authorship and immense collaboration in conducting/writing of the research.

REFERENCES

Al-Akhras, N. M. (2006). Durability of metakaolin concrete to sulfate attack. *Cement and Concrete Research*, *36*(9), 1727–1734. doi:10.1016/j.cemconres.2006.03.026

Al-Majidi, M. H., Lampropoulos, A., Cundy, A., & Meikle, S. (2016). Development of geopolymer mortar under ambient temperature for in situ applications. *Construction & Building Materials*, *120*, 198–211. doi:10.1016/j.conbuildmat.2016.05.085

Alanazi, H., Yang, M., Zhang, D., & Gao, Z. J. (2016). Bond strength of PCC pavement repairs using metakaolin-based geopolymer mortar. *Cement and Concrete Composites*, *65*, 75–82. doi:10.1016/j.cemconcomp.2015.10.009

Ambroise, J., Maximilien, S., & Pera, J. (1994). Properties of metakaolin blended cements. *Advanced Cement Based Materials*, *1*(4), 161–168. doi:10.1016/1065-7355(94)90007-8

Arellano-Aguilar, R., Burciaga-Díaz, O., Gorokhovsky, A., & Escalante-García, J. I. (2014). Geopolymer mortars based on a low grade metakaolin: Effects of the chemical composition, temperature and aggregate: binder ratio. *Construction & Building Materials*, *50*, 642–648. doi:10.1016/j.conbuildmat.2013.10.023

ASTM. C. 1991. 882-91–Standard test method for bond strength of epoxy resin system used with concrete by slant shear. American Society for Testing Materials, ASTM.

ASTM. C. 1999. 618 (1993) Standard specification for coal fly ash and raw or calcined natural pozzolan for use as a mineral admixture in concrete. Annual Book of ASTM Standards, Philadelphia, USA.

Barbosa, V. F., & MacKenzie, K. J. (2003). Synthesis and thermal behaviour of potassium sialate geopolymers. *Materials Letters, 57*(9), 1477–1482. doi:10.1016/S0167-577X(02)01009-1

Chen, J.-H., Huang, J.-S., & Chang, Y.-W. (2011). Use of reservoir sludge as a partial replacement of metakaolin in the production of geopolymers. *Cement and Concrete Composites, 33*(5), 602–610. doi:10.1016/j.cemconcomp.2011.03.002

Chen, J. Y., Lei, X. R., & Tao, S. (2014). *Simultaneous Reduction and Immobilization of Cr2O72-in Metakaolinite-Based Geopolymer Complexed Reductant.* Paper presented at the Advanced Materials Research.

Chindaprasirt, P., De Silva, P., Sagoe-Crentsil, K., & Hanjitsuwan, S. (2012). Effect of SiO_2 and Al_2O_3 on the setting and hardening of high calcium fly ash-based geopolymer systems. *Journal of Materials Science, 47*(12), 4876–4883. doi:10.100710853-012-6353-y

Cyr, M., & Pouhet, R. (2016). Carbonation in the pore solution of metakaolin-based geopolymer. *Cement and Concrete Research, 88*, 227–235. doi:10.1016/j.cemconres.2016.05.008

Davidovits, J. (2008). Geopolymer Chemistry and Applications. Saint-Quentin, France: Geopolymer Institute.

Duan, P., Yan, C., Zhou, W., Luo, W., & Shen, C. (2015). An investigation of the microstructure and durability of a fluidized bed fly ash–metakaolin geopolymer after heat and acid exposure. *Materials & Design, 74*, 125–137. doi:10.1016/j.matdes.2015.03.009

Duxson, P., Fernández-Jiménez, A., Provis, J. L., Lukey, G. C., Palomo, A., & Van Deventer, J. (2007). Geopolymer technology: The current state of the art. *Journal of Materials Science, 42*(9), 2917–2933. doi:10.100710853-006-0637-z

Gordon, L. E., San Nicolas, R., & Provis, J. L. (2014). Chemical characterisation of metakaolin and fly ash based geopolymers during exposure to solvents used in carbon capture. *International Journal of Greenhouse Gas Control, 27*, 255–266. doi:10.1016/j.ijggc.2014.06.005

Görhan, G., Aslaner, R., & Şinik, O. (2016). The effect of curing on the properties of metakaolin and fly ash-based geopolymer paste. *Composites. Part B, Engineering*, *97*, 329–335. doi:10.1016/j.compositesb.2016.05.019

He, P., Jia, D., Lin, T., Wang, M., & Zhou, Y. (2010). Effects of high-temperature heat treatment on the mechanical properties of unidirectional carbon fiber reinforced geopolymer composites. *Ceramics International*, *36*(4), 1447–1453. doi:10.1016/j.ceramint.2010.02.012

Huseien, G. F., Mirza, J., Ismail, M., Ghoshal, S., & Ariffin, M. A. M. (2016a). (in press). Effect of metakaolin replaced granulated blast furnace slag on fresh and early strength properties of geopolymer mortar. *Ain Shams Engineering Journal*.

Huseien, G. F., Mirza, J., Ismail, M., Ghoshal, S., & Hussein, A. A. (2017). Geopolymer mortars as sustainable repair material: A comprehensive review. *Renewable & Sustainable Energy Reviews*, *80*, 54–74. doi:10.1016/j.rser.2017.05.076

Huseien, G. F., Mirza, J., Ismail, M., & Hussin, M. W. (2016b). Influence of different curing temperatures and alkali activators on properties of GBFS geopolymer mortars containing fly ash and palm-oil fuel ash. *Construction & Building Materials*, *125*, 1229–1240. doi:10.1016/j.conbuildmat.2016.08.153

Ismaïl, M., Yusuf, T. O., Noruzman, A. H., & Hassan, I. (2013). *Early strength characteristics of palm oil fuel ash and metakaolin blended geopolymer mortar.* Paper presented at the Advanced Materials Research. 10.4028/www.scientific.net/AMR.690-693.1045

Jin, M., Zheng, Z., Sun, Y., Chen, L., & Jin, Z. (2016). Resistance of metakaolin-MSWI fly ash based geopolymer to acid and alkaline environments. *Journal of Non-Crystalline Solids*, *450*, 116–122. doi:10.1016/j.jnoncrysol.2016.07.036

Kong, D. L., Sanjayan, J. G., & Sagoe-Crentsil, K. (2007). Comparative performance of geopolymers made with metakaolin and fly ash after exposure to elevated temperatures. *Cement and Concrete Research*, *37*(12), 1583–1589. doi:10.1016/j.cemconres.2007.08.021

Kumar, S., Kumar, R., & Mehrotra, S. (2010). Influence of granulated blast furnace slag on the reaction, structure and properties of fly ash based geopolymer. *Journal of Materials Science*, *45*(3), 607–615. doi:10.100710853-009-3934-5

Lee, N., Kim, E., & Lee, H. (2016). Mechanical properties and setting characteristics of geopolymer mortar using styrene-butadiene (SB) latex. *Construction & Building Materials*, *113*, 264–272. doi:10.1016/j.conbuildmat.2016.03.055

Malhotra, V. M., & Mehta, P. K. (1996). *Pozzolanic and Cementitious Materials* (Vol. 1). Taylor & Francis.

Pacheco-Torgal, F., Abdollahnejad, Z., Camões, A., Jamshidi, M., & Ding, Y. (2012). Durability of alkali-activated binders: A clear advantage over Portland cement or an unproven issue? *Construction & Building Materials*, *30*, 400–405. doi:10.1016/j.conbuildmat.2011.12.017

Pacheco-Torgal, F., Castro-Gomes, J., & Jalali, S. (2008). Alkali-activated binders: A review: Part 1. Historical background, terminology, reaction mechanisms and hydration products. *Construction & Building Materials*, *22*(7), 1305–1314. doi:10.1016/j.conbuildmat.2007.10.015

Pan, Z., Sanjayan, J. G., & Collins, F. (2014). Effect of transient creep on compressive strength of geopolymer concrete for elevated temperature exposure. *Cement and Concrete Research*, *56*, 182–189. doi:10.1016/j.cemconres.2013.11.014

Phoo-ngernkham, T., Maegawa, A., Mishima, N., Hatanaka, S., & Chindaprasirt, P. (2015). Effects of sodium hydroxide and sodium silicate solutions on compressive and shear bond strengths of FA–GBFS geopolymer. *Construction & Building Materials*, *91*, 1–8. doi:10.1016/j.conbuildmat.2015.05.001

Poon, C.-S., Lam, L., Kou, S., Wong, Y.-L., & Wong, R. (2001). Rate of pozzolanic reaction of metakaolin in high-performance cement pastes. *Cement and Concrete Research*, *31*(9), 1301–1306. doi:10.1016/S0008-8846(01)00581-6

Ranjbar, N., Mehrali, M., Alengaram, U. J., Metselaar, H. S. C., & Jumaat, M. Z. (2014). Compressive strength and microstructural analysis of fly ash/palm oil fuel ash based geopolymer mortar under elevated temperatures. *Construction & Building Materials*, *65*, 114–121. doi:10.1016/j.conbuildmat.2014.04.064

Reig Cerdá, L., Borrachero, M., Istuque, D., Moraes, J., Akasaki, J., & Soriano, L. (2016). Behaviour of metakaolin-based geopolymers incorporating sewage sludge ash (SSA). *Materials Letters*, *180*, 192–195. doi:10.1016/j.matlet.2016.05.137

Rovnaník, P. (2010). Effect of curing temperature on the development of hard structure of metakaolin-based geopolymer. *Construction & Building Materials*, *24*(7), 1176–1183. doi:10.1016/j.conbuildmat.2009.12.023

Sabir, B., Wild, S., & Bai, J. (2001). Metakaolin and calcined clays as pozzolans for concrete: A review. *Cement and Concrete Composites*, *23*(6), 441–454. doi:10.1016/S0958-9465(00)00092-5

Sofi, M., Van Deventer, J., Mendis, P., & Lukey, G. (2007). Engineering properties of inorganic polymer concretes (IPCs). *Cement and Concrete Research*, *37*(2), 251–257. doi:10.1016/j.cemconres.2006.10.008

Sugama, T., Brothers, L., & Van de Putte, T. (2005). Acid-resistant cements for geothermal wells: Sodium silicate activated slag/fly ash blends. *Advances in Cement Research*, *17*(2), 65–75. doi:10.1680/adcr.2005.17.2.65

Tafraoui, A., Escadeillas, G., Lebaili, S., & Vidal, T. (2009). Metakaolin in the formulation of UHPC. *Construction & Building Materials*, *23*(2), 669–674. doi:10.1016/j.conbuildmat.2008.02.018

Tchakouté, H. K., Rüscher, C. H., Kong, S., Kamseu, E., & Leonelli, C. (2016). Geopolymer binders from metakaolin using sodium waterglass from waste glass and rice husk ash as alternative activators: A comparative study. *Construction & Building Materials*, *114*, 276–289. doi:10.1016/j.conbuildmat.2016.03.184

Wild, S., & Khatib, J. (1997). Portlandite consumption in metakaolin cement pastes and mortars. *Cement and Concrete Research*, *27*(1), 137–146. doi:10.1016/S0008-8846(96)00187-1

Zhang, Z., Provis, J. L., Reid, A., & Wang, H. (2014). Geopolymer foam concrete: An emerging material for sustainable construction. *Construction & Building Materials*, *56*, 113–127. doi:10.1016/j.conbuildmat.2014.01.081

Zhang, Z., Zhu, H., Zhou, C., & Wang, H. (2016). Geopolymer from kaolin in China: An overview. *Applied Clay Science*, *119*, 31–41. doi:10.1016/j.clay.2015.04.023

Zhou, C. H., & Keeling, J. (2013). Fundamental and applied research on clay minerals: From climate and environment to nanotechnology. *Applied Clay Science*, *74*, 3–9. doi:10.1016/j.clay.2013.02.013

Živica, V., Balkovic, S., & Drabik, M. (2011). Properties of metakaolin geopolymer hardened paste prepared by high-pressure compaction. *Construction & Building Materials*, *25*(5), 2206–2213. doi:10.1016/j.conbuildmat.2010.11.004

Chapter 5
Versatility of Cockle Shell in Concrete:
A Conspectus

ABSTRACT

A conspectus based upon a compelling topic, namely, versatility of cockle shell use in concrete to replace partially the natural coarse aggregates and river sand, which is yet to be investigated, is covered in this chapter. An introduction to enlighten the reader with this promising waste material precedes a review of environmental issues with cockle shell which would reduce harm to environment and preserve natural materials for future generation. Cockle trade is an important subtitle that covers cockle shell waste generation, research, and development related to the deployment on the use of cockle shell, processing cockle shell for making construction material are discussed in detail. Experiments were conducted, and the test data revealed that the use of cockle shell as partial replacement of coarse aggregates enhanced the strength of concrete and as partial replacement of sand improved the performance of mortar bricks.

INTRODUCTION

Construction industry contributes to the establishment of various types of structures in meeting with the demands of expanding population. In relation to that, the growing needs of construction trade requires natural resources such as granite aggregates and sand to be harvested from the environment.

DOI: 10.4018/978-1-5225-8325-7.ch005

Both granite aggregates extracted from local quarries and sand mined from river are utilized for the production of construction materials such as concretes and bricks. The continuous quarrying of granite aggregate, a non-renewable material, poses negative impact on the environment in terms of destruction of wild life habitat, climate change and possible depletion in future. The excessive river sand mining has opened up the door for water pollution and ecological imbalance to the river bed environment which directly affects the quality life of surrounding community. In view of sustainable construction, explorations for other alternative materials which could be used as partial or full replacement of coarse aggregates or sand would not only benefit the construction industry and also contribute to the cleaner environment.

At the same time, the active cockle trade in Malaysia generates a large amount of cockle shells that are disposed-off as waste. Cockle, being an important protein source in South East Asian region contributes to the growth of this industry. In practice, the harvested cockle are processed to obtain its edible meat before being produced as canned food or sent to fresh market. The cockle shells are usually dumped as waste in large quantity at dumping site which causes pollution. Issues of waste disposal at landfill sites become a burden to those industries owing to the extra cost which needs to be spent for waste management. In addition, disposal of these wastes in increasing quantity is creating a negative impact on the environment. It is observed that the freely available wastes when used as mixing ingredient for the production of construction material would produce an environmental friendly material. This would decrease the dependency on natural resources and also reduces the amounts of cockle trade waste ending at landfill.

Related Environmental Issues

The rapidly growing Malaysian construction industry has increased the demand for concrete production, which directly leads to higher consumption of granite aggregates and fine aggregates. Although, this country still have sufficient granite aggregates reserves but continuous quarrying activities would cause ecological imbalance, affect community's healthy life style and aggregates depletion in the future. This issue has been touched by local researchers (Hainin et al. 2012) who stated that extensive use of granite aggregates in construction would disturb the environment and eventually bring the local granite supply to an end. Imported aggregates have been used increasingly over the years (Malaysian Geoscience and Mineral Department, 2011). Continuous procurement of the imported aggregate in larger amount as production of natural aggregate supply is insufficient to meet the increasing demand of building industry which would escalate the concrete materials' cost. Unless, actions are taken to reduce the high dependency on granite aggregates by using other types of material as alternative aggregates, the depletion of aggregates is inevitable in future. In short, smart approach of integrating agricultural wastes as aggregates in concrete would save energy, conserve natural resources and also reduce construction materials cost (Prusty and Patro, 2015).

The sand mining activities at the rivers are becoming more active in order to meet the construction industry need. This has resulted in negative impact towards river environment in terms of water quality and destruction of aquatic life habitat. Sand mining imposes negative impact on the environment in terms of reduced water quality, destabilization of stream bed and bank which in turn cause the destruction of riverine vegetation (Asyraf et al., 2011) thus leading to ecological imbalance. Other than that, there are many countries facing sand supply shortage (Rashad, 2016). In order to reduce high dependency of the construction industry on sand supply and ensure the sustainability of river environment, one of the alternatives is to use the freely available solid waste to replace the river sand partially or perhaps fully to produce concrete.

At the same time, the active cockle trade contributes to the positive economy development of the local community to meet the increasing demand of consumers. This fishery industry generates a large quantity of cockle shells which are disposed at dumping site. According to Boey et al. (2011), the active cockle trade has led towards the generation of abundant waste shells. The cockle's production is increasing which increased the retail value of cockles alone by RM91.60 million in 2010 from 68.60 million the previous year, i. e., an increase of 33.53% (Department of Fisheries, 2010). It is expected that the availability of cockle shells as waste would be in larger amount as well which in turn will pose negative impact to the nearby area. Thus, concerns towards preserving the cleanliness of the environment and reducing amount of waste dumped at landfills has motivated the present research to investigate the possibility of integrating cockle shells as mixing ingredient in the production of concrete and bricks. Existence of environmental friendly construction materials would contribute towards a more sustainable construction and well balanced ecosystem.

Cockle Trade

Cockles or "*Anadara Granosa*" as shown in Figure 1 is a type of shellfish that grows in muddy coastal area. It is quite common to use it in local dishes owing to its cheap price. It began to commercialize in Malaysia by the end of 20th century. Cockle breeding are usually carried out at the mangrove swamps on west coast of Peninsular Malaysia from Kedah coast to the north of Johor which is located in the south of Malaysia. The cockle aquaculture areas in Malaysia extend about 10,383.09 hectares contributing a production of 78,024.7 tons in year 2010 (Department of Fisheries, 2010). The cockles which are normally produced in form of processed food or fresh edible meat without shell are sold to consumer inside and outside the country. This industry continues to grow as shown in Figure 2, as the demand from consumer increases.

Cockle Shell Waste Generation

As the production of cockle increases, the by-product from this industry increases as disposal waste material. This fact has been highlighted by Mijan et al. (2011) who stated that cockle shell is abundantly available in Malaysia

Figure 1. Cockles

Figure 2. Production of cockle shell in Malaysia (2007-2010)
Source: Department of Fisheries (2007-2010)

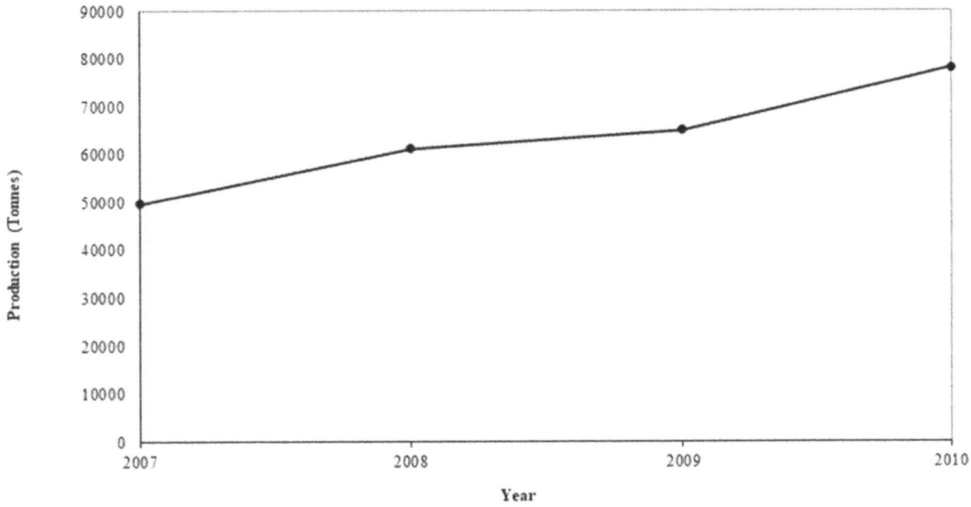

as a by-product from seafood industry. Large amount of cockle shells have created the problem to the environment and community. The cockle shells that have been dumped and left untreated will cause an unpleasant smell and disturbing view to the surrounding (Rashidi et al., 2011). The shell is hard which makes it difficult to be disposed as compared to other biodegradable materials such as coconut shell or oil palm shell. Leaving this waste at dumping site to biodegrade is not the best approach to reduce the use of landfill space. Processing this shell waste to be used as a construction material would reduce the amount of waste disposed at landfill and save the landfill space. Figure 3 illustrates the cockle shell generation process beginning from the cockles brought by the fisherman boat to the factory until the shell is produced. Figure 4 shows the dumping site of the cockle shells.

Research and Development on the Use of Cockle Shell

The beauty of cockle shell (see Figure 5) has attracted the local community in the fishing village to produce handcraft by using it. However, its use in the handcraft industry is limited and the production of cockle shell is in abundance allowing it to be used for other applications. In chemical industry field, the local researchers discovered its potential for use as a catalyst in biodiesel production from palm olein. The high calcium oxide content in cockle shell

Figure 3. Cockles are processed at the factory and its shell thrown as waste

Figure 4. Cockle shells are disposed at dumping site

Figure 5. Cockle shells are washed and dried before ready to be used

(Table 1) has made it suitable to repair bones in the medical field. Being a natural material, cockle shell has also been found useful in fabrication of artificial reef which is more environmental friendly to the marine. So far, very limited research is available on the use of cockle shell in the production of construction materials, namely concrete and bricks, which are widely used in the country. Only one study was found in Thailand where cockle shells were reported as a replacement of natural aggregates (Tonnayopas et al.,

Table 1. Chemical composition of cockle shell

Chemical Composition	(%)
Calcium Oxide (CaO)	67.28
Sodium Oxide (Na_2O)	0.50
Iron Oxide (Fe_2O_3)	0.40
Aluminium Oxide (Al_2O_3)	0.27
Strontium Oxide (SrO)	0.19
Sulphur Trioxide (SO_3)	0.19
Magnesium Oxide (MgO)	0.14
Phosphorus Pentoxide (P_2O_5)	0.07
Potassium Oxide (K_2O)	0.04
Chlorine (Cl)	0.03
Manganese Oxide (MnO)	0.02
Titanium Oxide (TiO_2)	0.02

2011). This research study used blended cockle shells with rice husk ash. Contrary to this, our research incorporated only cockle shells in the concrete. Therefore, this research study was conducted to explore the potential of this waste material as mixing ingredient for the production of concrete and bricks.

Processing Cockle Shells as a Construction Material

This research investigated the potential of cockle shell as partial replacements of coarse aggregates and sand. All the shells collected from the cockle processing factories or dumping sites were washed thoroughly using tap water to remove clay, dust, sand, debris, etc., followed by air dried as shown in Figure 6. They were then ready for use as partial replacement of coarse aggregates in concrete. For the partial replacement of fine aggregates, the shells were crushed using jaw crusher and sieved by passing through 1.18 mm sieve. Figure 7 illustrates the image of cockle shells as partial replacements of coarse and fine aggregates.

Partial Coarse Aggregates Replacement

The present research investigated the effects of integrating various percentages of cockle shells as partial replacement of coarse aggregates (0%, 5%, 10%, 15%, 20%, 25% and 30%) towards workability and compressive strength of concrete. Based on the results (Figure 8), the concrete workability reduced as the percentage of cockle shell increased. As the amount of cockle shells

Figure 6. Cockle shells are washed and dried before ready to be used

Figure 7. Partial coarse aggregates replacement and fine aggregates replacement

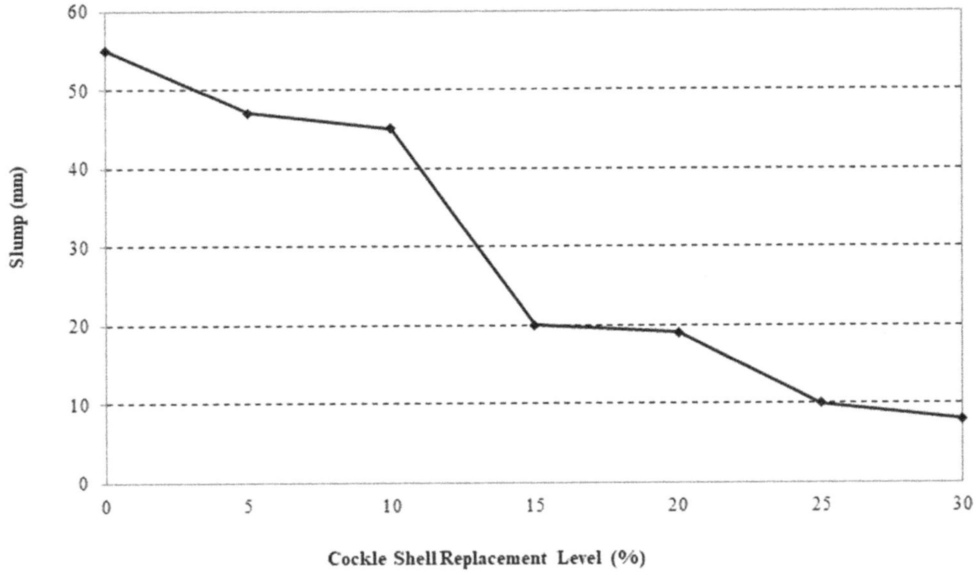

Figure 8. Effect of cockle shell on workability of concrete

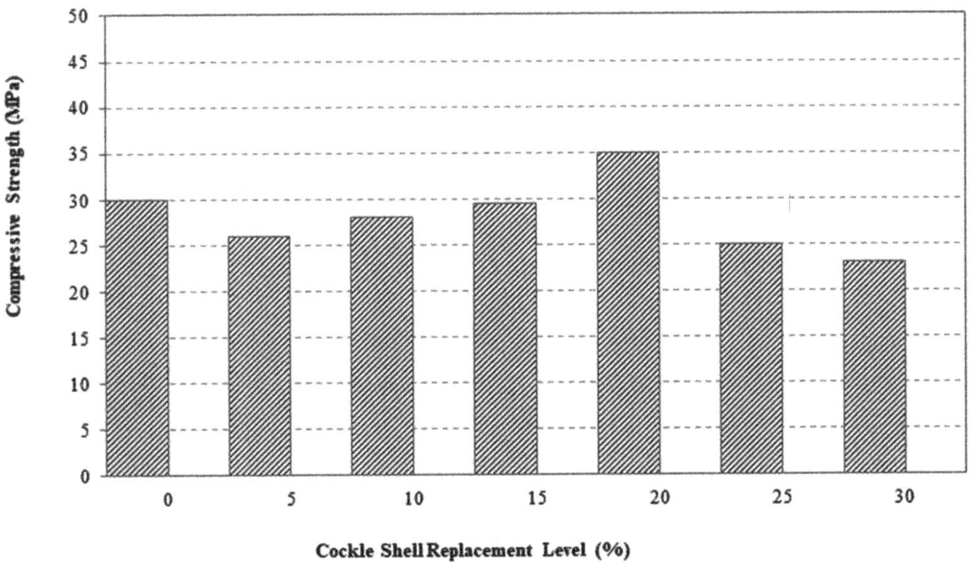

increased, their rough textured surface caused the mix to become more difficult to be mixed which led to lower the slump value. Concrete mixture containing the highest content of cockle shells (30%) showed the least workability. This could probably be credited to the variation in cockle shells texture which are rougher (due to the symmetrically radial ribs running on the outer shell) than the natural aggregates (Muthusamy and Sabri, 2012). The rougher surface creates more friction thus decreasing the fluidity of the mix containing higher percentage of cockle shells. These results are in line with the findings of Mindess et al. (2003) who highlighted that the use of smooth particles for concrete mixing would produce more workable mixture compared with the ones consisting rougher particles.

Furthermore, the difference in the shape of angular granite as compared to the curved and roughly parabolic shape cockle shells was found to affect the concrete workability. Past researcher (Adewuyi and Adegoke, 2008) highlighted that the workability of concrete can be affected by the shape of coarse aggregates. Other than that, the use of solid granite which is heavier and replaced by weight, the specific surface area increases with the increment of cockle shell contents in the mix. This situation requires a larger quantity of cement paste than the amount available in the mix to coat the cockle shell particles. This, in turn, leads to lesser workability for concrete mix containing larger percentage of cockle shells as partial replacement of coarse aggregates.

In terms of strength performance, the partial replacement of granite aggregates with cockle shells enhanced the strength performance of concrete when a right mix proportion was formulated. Mix consisting 20% replacement of cockle shells exhibited the highest value of compressive strength which outshined other mixes including plain concrete as shown in Figure 9. The surface texture of cockle shells which are rougher than granite aggregates improve the bonding and increase the inter particle friction, which, in turn, enhances the compressive strength of concrete. Mindess et al. (2003) highlighted that the aggregates with rough textured surfaces will improve the mechanical component of the bond.

On the other hand, it is apparent that too much of cockle shells tend to decrease the compressive strength of concrete as can be seen in the performance of concrete mix containing 30% cockle shells. It could be justified since too much cockle shells (which means higher effective surface area) would lead to insufficient proportion of cement paste thus produce poor bonding properties of the matrix with aggregates. In addition, the reduction in concrete workability with an increase of cockle shells makes the concrete mix difficult to be compacted thus leading to higher porosity or exhibiting lower strength.

Figure 9. Compressive strength of concrete containing cockle shell as partial course aggregates replacement at 28 days

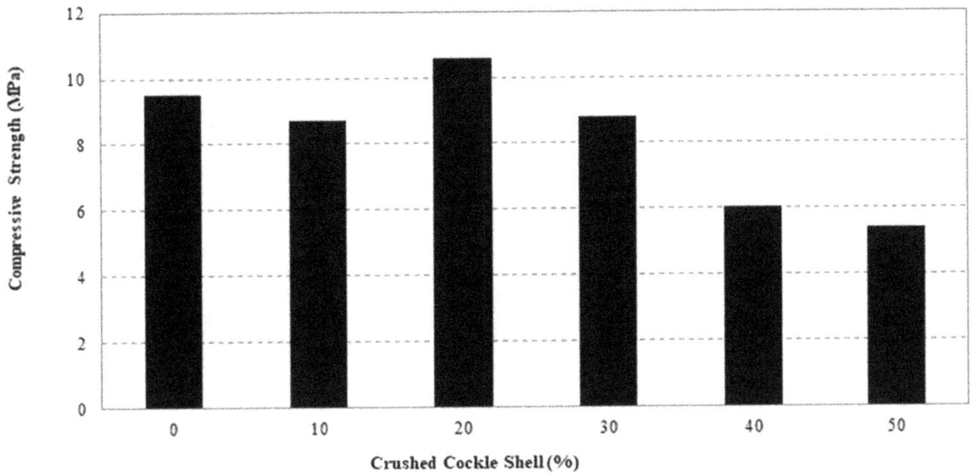

Overall, this study indicated that the optimum replacement of cockle shells is possible to produce a workable mix for concrete work.

Partial Sand Replacement

The potential use of cockle shells as partial sand replacement is discussed in this section. Based on the results of investigation conducted by adding crushed cockle shells as partial replacement of fine aggregates at varying percentages (0% - 50%) in mortar, it was found that the inclusion of crushed cockle shells had influenced the compressive strength performance of the mix. As depicted in Figure 10, the highest compressive strength value was achieved by adding 20% crushed cockle shells with only 80% of fine aggregates as natural river sand. This positive contribution could probably be due to the filling effect of this fine waste material which makes the concrete denser and stronger (Muthusamy et al., 2016). Although inclusion of more crushed cockle shells would reduce the larger amount of Natural River sand which makes the material more environmental friendly, the utilization of this waste material in the mix need to be limited. This is because too much inclusion of crushed cockle shells in the mix causes the compressive strength of the mix to drop. It was observed that beyond 20% replacement of crushed cockle shells, the strength of mortar continues to decrease as their amount of integration become larger.

Figure 10. Compressive strength of mortar cube with various percentage of crushed cockle shell as partial sand replacement

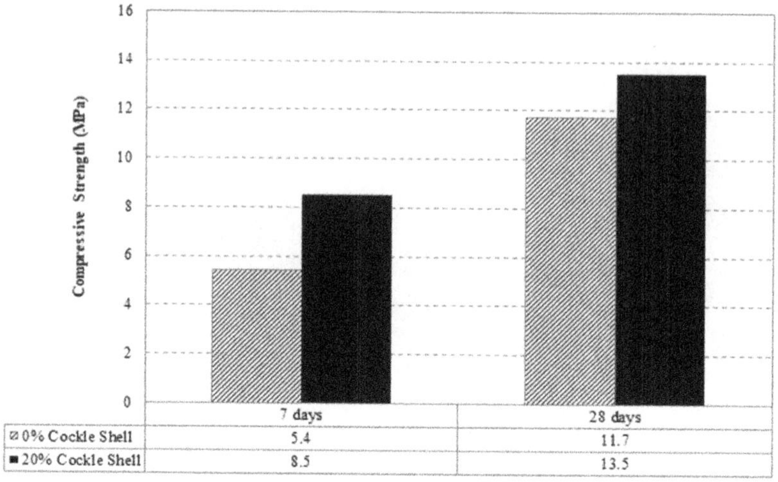

Looking at the strength performance of bricks produced using 100% Natural River sand and the one containing 20% cockle shells as illustrated in Figure 11; both specimens exhibited a continued strength increment throughout the curing period. However, bricks containing 20% cockle shells

Figure 11. Compressive strength of control specimen and brick containing 20% cockle shell at 7 and 28 days

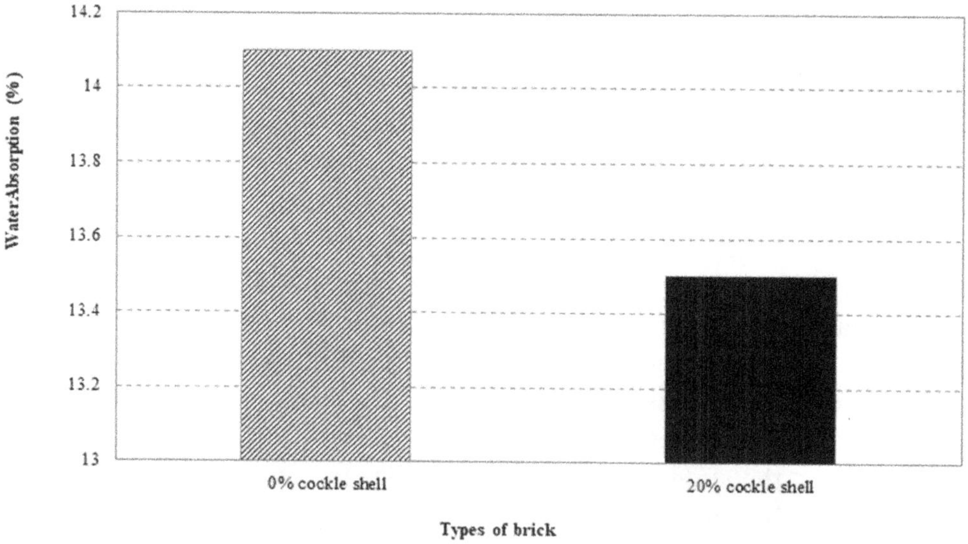

performed better in all curing ages. In addition, the moisture absorption of bricks was also lower than the control specimen as depicted in Figure 12. The use of crushed cockle shells as partial replacement of fine aggregates increased the denseness of concrete internal structure through filling effect which contributed to enhance the strength achievement and lower the moisture absorption. This finding encourages the use of cockle shells as partial sand replacement in cement sand bricks production which consumes lower quantity of river sand. Most importantly, the pollution issues related to the disposal of cockle shells as dumping site can be reduced as lesser amount of shells would end up as waste.

CONCLUSION

Based on the experimental investigation, the following remarks can be highlighted:

1. Cockle shell has the potential to be used as partial replacements of fine/ or coarse aggregates in concrete.
2. When cockle shells are used as partial replacement of coarse aggregates in concrete, addition of up to 20% contributed to its strength development.

Figure 12. Water absorption of control specimen and brick containing 20% cockle shell at 28 days

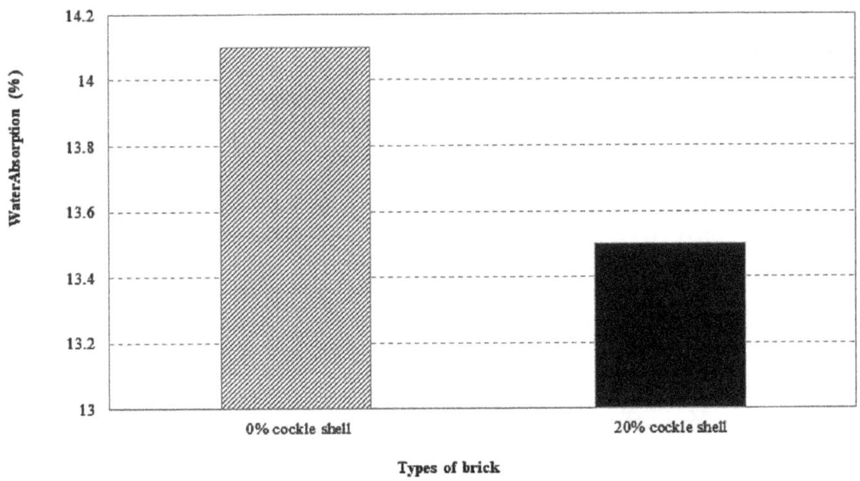

3. Integration of suitable contents of crushed cockle shells between 10 and 20% enhanced the compressive strength of concrete and cement sand bricks through filling effect.
4. Approaches taken to turn the cockle shells waste to wealth creation through production of environmental friendly construction material would not only reduce the high dependency of construction industry on natural resources but also minimize the amount of waste ending at dumpsite.

ACKNOWLEDGMENT

We, the editors greatly appreciate the exceptional contribution of Dr. Kharunisa Muthusamy for carrying out this research and writing of the chapter. Indeed, members of her team, Dr. Nur Hafizah Abd Khalid, Dr. Nur Farhayu Ariffin, Dr. Ali Umara Shettima, Dr. Ghasan Fahim Huseien, and Dr. Han Seung Lee deserve to be equally acknowledged for their co-authorship and immense collaboration in conducting/writing of the research.

REFERENCES

Adewuyi, A. P., & Adegoke, T. (2011). Exploratory study of periwinkle shells as coarse aggregate in concrete work. *Journal of Engineering and Applied Sciences (Asian Research Publishing Network)*, *3*(6), 1–5.

Asyraf, M. A., Maah, M. J., Yusoff, I., Wajid, A., & Mahmud, K. (2011). Sand mining, effects causes and concerns: A case study from Bestari Jaya Selangor Peninsula Malaysia. *Scientific Research and Essays*, *6*(6), 1216–1231.

Boey, P. L., Maniam, G. P., Abd Hamid, S., & Ali, D. M. H. (2011). Utilization of waste cockle shell (Anadara granosa) in biodiesel production from palm olein: Optimization using response surface methodology. *Fuel*, *90*(7), 2353–2358. doi:10.1016/j.fuel.2011.03.002

Department of Fisheries Malaysia. (2010). *Annual Fisheries Statistic 2010* (Jabatan Perikanan Malaysia, 2010). Author.

Hainin, M. R., Yusoff, N. I. M., Sabri, M. F. M., Aziz, M. A., Hameed, M. A. S., & Reshi, W. F. (2012). Steel slag as an aggregate replacement in Malaysian hot mix asphalt. *International Scholarly Research Network*, 1-5.

Malaysian Geoscience and Mineral Department. (2011). Aggregate production in Malaysia. Technical Report, Malaysia.

Mijan, N. A., Sahari, F., & Yusoff, S. B. (2011). Cockle shell in artificial reef construction. *Proceedings of International Conference on Creativity and Innovation for Sustainable Development*.

Mindess, S., Young, J. F., & Darwin, D. (2003). *Concrete*. Prentice Hall.

Muthusamy, K., & Sabri, N. A. (2012). Cockle shell: A potential partial coarse aggregate replacement in concrete. *International Journal of Science. Environmental Technology*, *1*(4), 260–267.

Muthusamy, K., Tukimat, N. N. A., Sarbini, N. N., & Zamri, N. A. (2012). Exploratory study on the use of crushed cockle shell as partial sand replacement in concrete. *International Journal of Research in Engineering and Science*, *4*(2), 67–71.

Prusty, J. K., & Pastro, S. K. (2016). Properties of fresh and hardened concrete using agro wastes as partial replacement of coarse aggregate - A review. *Construction & Building Materials*, *82*, 101–113. doi:10.1016/j.conbuildmat.2015.02.063

Rashad, A. (2016). Cementitious materials and agricultural wastes as natural fine aggregate replacement in conventional mortar and concrete. *Journal of Building Engineering*, *5*, 119–141. doi:10.1016/j.jobe.2015.11.011

Rashidi, N. A., Mohamed, M., & Yusup, S. (2011). A study of calcination and carbonation of cockle shell. *International of Chemical, Molecular, Nuclear Materials and Metallurgical Engineering Journal*, *512*, 1118–1123.

Chapter 6
Durability of Mortars Containing Ceramic Tile Waste Exposed to Sulphate Attack

ABSTRACT

Chapter 6 deals with the durability performance of mortar containing ceramic tile waste exposed to sulphate attack. The introduction discusses the latest development regarding this subject as no case study has been found where ceramic tile waste was actually used in the field. This study investigates the sulphate resistance of ceramic mortar by using sulphate solution and tested the visual appearance of specimens, mass loss, residual compressive strength, and microstructure analysis up to 18 months of sulphate exposure. The ceramic mortar demonstrated superior advantages with respect to visual appearance and mass change with low values of strength loss upon exposure to sulphate solutions. Therefore, ceramic waste in the form of fine aggregates and fine powder can be used in mortar production with comparable strength and improvement in the fresh and hardened state properties of the mortar as compared with the OPC mortar.

INTRODUCTION

Limited research was carried out on the blended cement mortars incorporating ceramic powder exposed to sulphate attack. Generally aggressive chemicals such as sodium sulphate (Na_2SO_4), magnesium sulphate ($MgSO_4$) and calcium sulphate ($CaSO_4$) which are available in some soils or groundwater will attack

DOI: 10.4018/978-1-5225-8325-7.ch006

cement paste in concrete or mortar. These chemicals will degrade and/or deteriorate the cement paste. In soil, these aggressive agents are in the form of salts where, when dissolved in water will react with $Ca(OH)_2$ present in ordinary Portland cement paste. According to periodic table of elements, sodium ions (Na^{+1}) are more reactive than calcium ions (Ca^{+2}) and can be more harmful to cement paste. The reaction between these aggressive agents and cement paste cause an expansion and swelling which will consequently crack the concrete or mortar. One of the main issues related to concrete or mortar is the attack by sulphate ions (SO_4^{-2}) which would decrease the compressive strength over a long period of time. The decrease in compressive strength of concrete in infrastructures can be a critical issue related to serviceability of structure. Series of researches have been conducted to fully understand the mechanism of sulphate ions (SO_4^{-2}) reactions (Bhutta et al., 2013). According to the findings reported by Mehta and Monteiro (1998), the reaction of sulphate ions (SO_4^{-2}) with cement paste of concrete occurred in two phases. In the primary phase, the sulphate ions (SO_4^{-2}) react with calcium hydroxide ($Ca(OH)_2$) and produce gypsum ($CaSO_4.2HO_2$) which fills the pores in concrete or mortar. In the secondary phase, the concentration of sulphate ions (SO_4^{-2}) increases and results in the transformation of mono-sulpho-aluminate into long needle-shaped crystals of ettringite ($3CaO. Al_2O_3. 3CaSO_4{\cdot}32H_2O$). The ettringite causes expansion in concrete or mortar which consequently reduces the compressive strength. In fact, not only sulphate ions (SO_4^{-2}) in the solution reacts with calcium hydroxide ($Ca(OH)_2$), other ions like magnesium ions (Mg^{+2}) also react with cement components and cause additional deterioration to concrete structures.

Usually, to improve the resistance of structures exposed to sulphate attack, pozzolanic materials are added in concrete or mortar mixtures. Extensive literature studies are available on the beneficial use of pozzolanic materials such as fly ash, palm oil fuel ash, rice husk ash, metakaolin, and silica fume to improve the resistance of concrete or mortar against sulphate attack (O'Farrell et al., 2006; Ramadhansyah et al., 2012b; Budiea et al., 2010; Jaturapitakkul et al., 2007). Such pozzolanic materials decrease the permeability of concrete and mortars by two mechanisms; firstly by reducing water requirements for mixture (using fly ash). Secondly, all pozzolanic materials contain high amount of amorphous silicate (SiO_2) which react with calcium hydroxide and produce calcium silicate hydrate. The calcium silicate hydrate gives denser structure to concrete which provide high resistance against sulphate attack.

Several research studies have been carried out to investigate the effect of ceramic waste on concrete's resistance to sulphate attack. Yunhong et al. (2014) observed that the sulphate corrosion resistance of concrete incorporating ceramic powder is better than the control concrete. Also with the increase in ceramic powder replacement level, the sulphate corrosion resistance of concrete showed a better performance. Similarly, Vejmelková et al. (2012) reported that by increasing the amount of ceramic powder up to 30% demonstrated better resistance to expansion which is similar to the studies conducted by Toledo et al. (2007) with 40% ceramic powder as cement replacement and up to 30% in work by O'Farrell et al. (2006). All these could be attributed to greater pozzolanic activity. Besides, Binici et al. (2008) reported the effect of utilizing marble aggregates as coarse aggregates replacement together with river sand and ground blast furnace slag as fine aggregates on sulphate resistance of concrete. However, our literature survey study showed no case studies where this waste material (ceramic) was used in the field. It was shown that the concrete containing marble aggregates had better resistance towards sulphate attack. Therefore, it is necessary to study the effect of concrete containing ceramic waste exposed to sulphate attack.

Sulphate Resistance

The sulphate resistance test was conducted to study the performance of ceramic mortar exposed to sulphate solution. The sulphate resistance of ceramic mortar specimens was evaluated based on the effect of sulphate solution on visual appearance of specimens, mass loss, residual compressive strength and microstructure analysis up to 18 months of sulphate exposure.

Visual Appearance

Figure 1 shows the visual appearance of the Ordinary Portland cement (OPC) and ceramic mortars after immersion in 5% sodium sulphate (Na_2SO_4) solution. The changes in dimension and condition of specimens were observed and recorded. As seen in Figure 1, there are no changes observed on the surface of the ceramic mortar, whereas for the OPC mortar, it shows significant surface deterioration after 18 months of exposure in 5% Na_2SO_4 solution. There was white sediment of gypsum observed in the crack formed because of sulphate attack on the OPC mortar. In contrast, mortar containing ceramic did not

Figure 1. Appearance of mortar cubes exposed to 5% NA2So4 solution for 18 months

show any change of dimension and remained structurally intact without any visible cracks.

Mass Change

The mass change of OPC and ceramic mortars after 18 months of immersion in 5% Na_2SO_4 solution is shown in Figure 2. After 3 months of immersion period, the mass of OPC mortar reduced by 0.7% while for ceramic mortar, it increased by 1.7% from its initial mass. Both the OPC and ceramic mortars showed an increment in mass at early age of immersion period. However, the OPC mortar tended to reduce the mass after 2 months of immersion period by 5.1% from its initial weight. This probably was due to the acceleration of sulphate attack which led to the formation of several products from the reaction between sodium sulphate and hydrates of cement, some of which expand within the specimens. This expansion, as it is very well known, is

Figure 2. Mass change of OPC and ceramic mortars in Na2So4

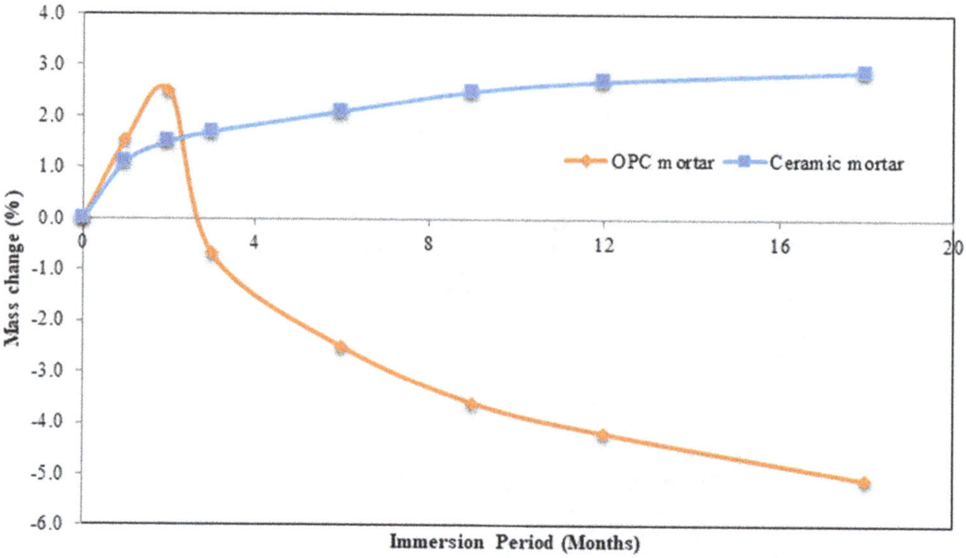

based on the formation of sulphate reaction products of ettringite and gypsum that leads to cracking.

Besides, the mass gain of ceramic mortar during the immersion period could be attributed by the sulphate reaction that increases the formation of gypsum and absorption of sulphate solution. This reaction is known to cause an increase in solid volume of the system, and may lead to the expansion of mortar. Furthermore, after 18 months of immersion, the ceramic mortar was still gaining weight while the OPC mortar already lost the weight due to the expansion and cracking. It is worth mentioning that the primary ettringite produced at early age does not produce cracking because the cement paste is still in the plastic state and, therefore, can accommodate any increase in volume. Similar finding was also reported by Ahmad and Noor (2011) by using different type of pozzolanic materials.

Residual Compressive Strength

Figure 3 shows the residual compressive strength of OPC and ceramic mortars after 18 months of immersion in Na_2SO_4 solution. After 18 months of immersion period, the OPC and ceramic mortars showed a reduction in compressive strength by 41.1% and 16.8%, respectively, from their initial

Figure 3. Residual compressive strength of OPC and ceramic mortars in Na2So4

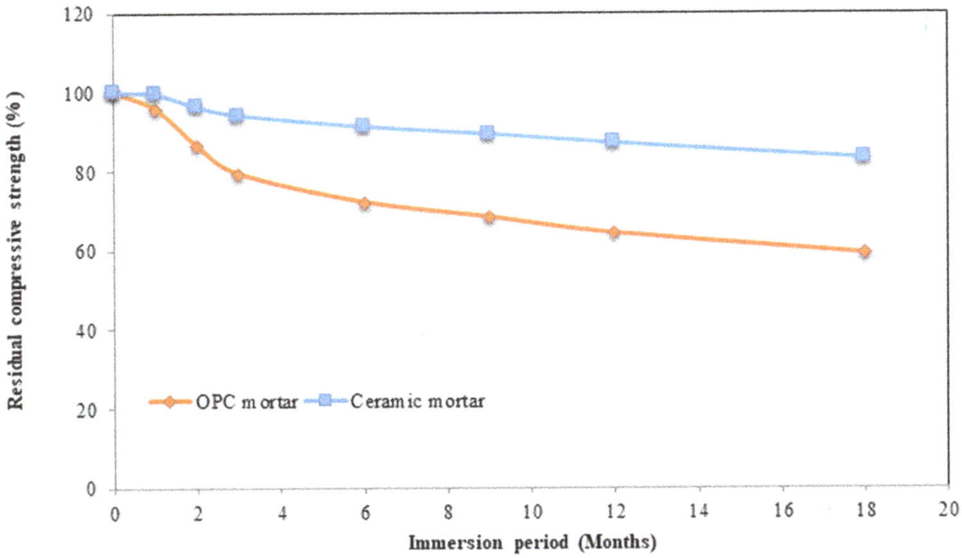

strength. It was observed that the compressive strength of mortar reduced with an increase in time of immersion. This is due to the reaction between sulphate and $Ca(OH)_2$ that led to the decomposition of cement hydration product that caused strength reduction. However, the ceramic mortar showed lower decrease in compressive strength compared with the OPC mortar. This is attributed to the gradual pozzolanic reaction between the ceramic powder and $Ca(OH)_2$ available thus, produced additional secondary C-S-H gel in the cement matrix. These additional hydrates increase the density of the matrix and refine the pore structure. This can be explained on the basis that the use of fine ceramic powder as a replacement of cement not only decreases the $Ca(OH)_2$ content, but also the unreacted fine ceramic powder will act as a micro filler. This reduces the voids between aggregates and binder which leads to a denser mortar.

In addition, the replacement of OPC with ceramic powder also dilutes the C_3A content, which reduce the formation of ettringite and thaumasite. These products expand within the mortar matrix, increase the internal stresses and eventually produce micro cracks. Therefore, the mortars lose their strength and produce cracks and deterioration. Besides, the amount of CaO presence in the binder materials also plays a significant role in the production of $Ca(OH)_2$. Thus, by using higher cement replacement with ceramic powder, the amount of $Ca(OH)_2$ is also reduced.

Figure 4. Relationship between compressive strength and immersion period

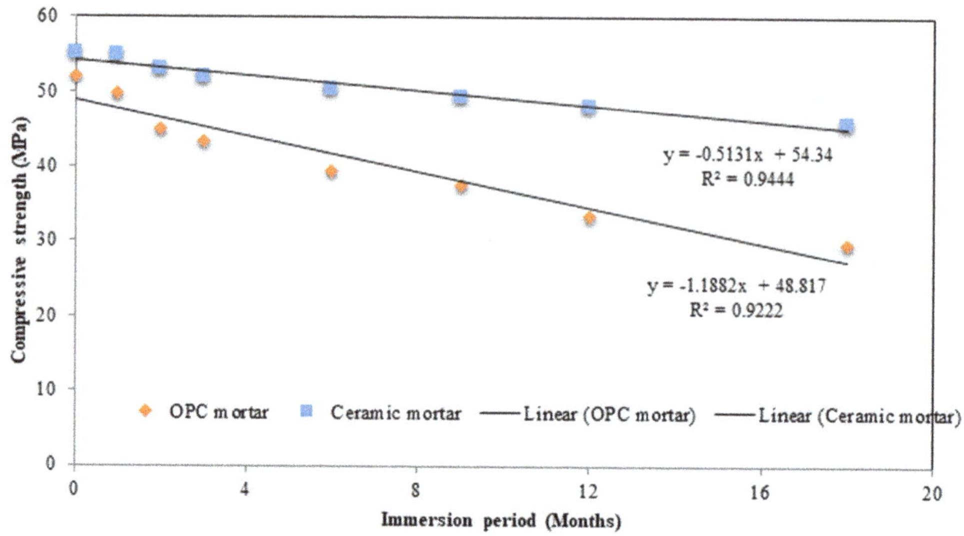

Figure 4 shows the relationship between compressive strength and immersion period in Na_2SO_4 solution of the OPC and ceramic mortars. It can be seen from the Figure 4 that the compressive strength of both types of mortars decreases with the increase in immersion period. The relationship between compressive strength and immersion period follows a linear pattern. The coefficient of determination (R^2) value of 92% and 94% for OPC and ceramic mortars, respectively, show a good correlation and prediction of the relationship.

Microstructure Analysis

The deterioration of mortar by the formation of gypsum goes through a process leading to a reduction in strength and followed by expansion. Gypsum formation is also known to be the first step of ettringite formation which can be considered as the major cause of deterioration due to sulphate attack. The Scanning electron microscopy (SEM) and Energy dispersive x-ray spectroscopy (EDX) images for both samples are shown in Figures 5 and 6, respectively. It can be seen that more needle crystals of ettringite were produced in the OPC mortar sample compared with ceramic mortar sample. This could be due to the amount of $Ca(OH)_2$ produced during the cement hydration process. The EDX in OPC mortar sample showed the existence of extended crystalline structures of gypsum and ettringite as shown by the

Figure 5. SEM and EDX images of OPC mortar in Na2So4 solution

Figure 6. SEM and EDX images of ceramic mortar in Na2So4 solution

presence of Ca, O, Al and S. However, different EDX spectrum can be seen from ceramic mortar sample which consisted of O, Si and Ca where more C-S-H gel was produced due to the gradual pozzolanic reaction. Therefore, it can be said that the ceramic mortar exhibited superior resistance than OPC mortar when exposed in sulphate solution.

The X-ray Diffraction (XRD) analysis of OPC and ceramic mortars samples, before and after 18 months of immersion in Na_2SO_4 solution, are shown in Figures 7 and 8. The XRD patterns for both samples showed similar results as shown in the SEM and EDX images. The high peak intensity of $Ca(OH)_2$

Figure 7. XRD pattern of OPC mortar sample after 18 months of immersion in Na2So4

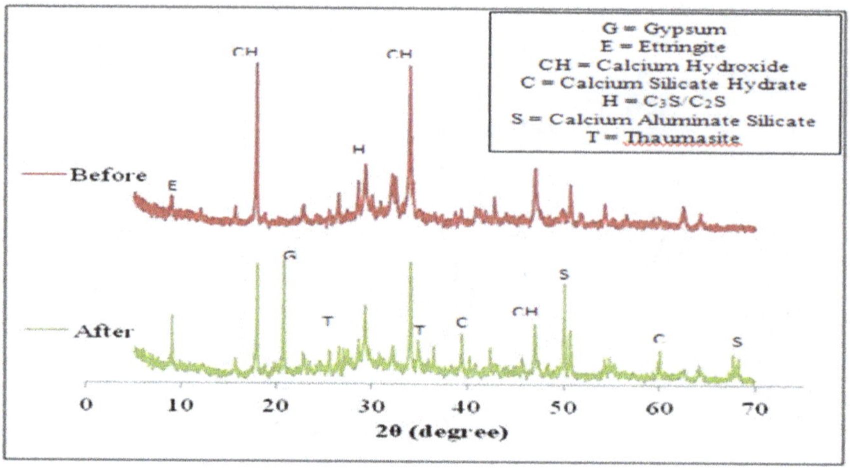

Figure 8. XRD pattern of ceramic mortar sample after 18 months immersion in Na2So4

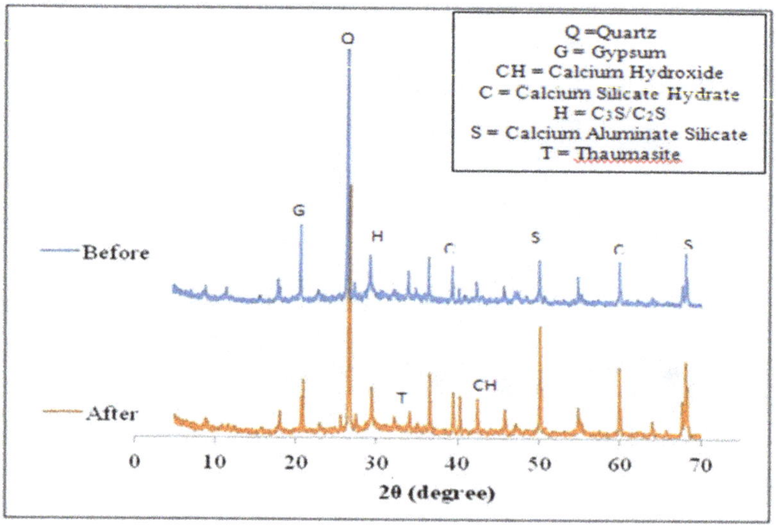

in OPC mortar sample before immersion produced a higher production of gypsum that led to the production of etttingite and thaumasite. Thaumasite can also be formed along with other calcium silicate hydrates (C-S-H) during cement alteration, especially when sulphate attack develops. The separation of peaks attributed to thaumasite and ettringite is greater at 2θ degree between 22θ degree and 29θ degree scale as illustrated in the Figures 7 and 8. The sulphate attack in mortar is generally attributed to the formation of expansive ettringite and gypsum. Therefore, the increase in peak intensity of this mineral shows the weak sulphate resistance and deforms the mortar. Due to less $Ca(OH)_2$ amount in the ceramic mortar sample, the amount of thaumasite and ettringite produced also reduced and high peak intensity of C-S-H gel can be observed at 39θ and 60θ scale. This is due to the reaction of the silica from ceramic powder with the excessive $Ca(OH)_2$ from cement hydration. Still high peak intensity of quartz is observed in the XRD analysis of ceramic mortar sample after 18 months of immersion due to the high amount of crystalline silica content that acted as micro filler and increased the packing effect of mortar.

The Fourier Transforms Infrared Spectroscopy (FTIR) spectra shown in Figures 9 and 10 indicate major bands at approximately 3645, 3450, 1645, 1425, 1110, 985, 470 cm^{-1} in the OPC mortar sample and 3450, 1645, 1425, 1015 and 463 cm^{-1} in ceramic mortar sample before immersion in the Na_2SO_4 solution. The structure of molecular water in the system is characterized by the O-H stretching band from $Ca(OH)_2$ decomposition, from 3200 to 3700 cm^{-1}, while the bending of the chemically bonded H-O-H was located at 1645 cm^{-1}, related to water bound in the hydrated products. Besides, the bands at 1005 to 1035 cm^{-1} are assigned to quartz as the crystalline phase in both samples. The FTIR spectra of ceramic mortar sample before and after immersion in Na_2SO_4 showed only minor differences. However, the reaction of the OPC mortar sample after exposure showed the decomposition of C-S-H and O-H phases in the microstructure. Figure 9 shows differences between the spectra obtained from the exposed and unexposed OPC mortar samples. The decomposition of the main binder, C-S-H gel is associated with the shifting to the new bands at 3645, 1015, 786 and 463 cm^{-1} in the sample after immersion. This shows that the OPC mortar is altered by the Na_2SO_4 solution exposure.

CONCLUSION

This study principally focused on evaluating the performance of mortar incorporating ceramic waste and exposed to sulphate attack. An OPC mortar, designed to have a 40 MPa compressive strength at 28 days was used as a reference. In terms of fresh state properties, the filling ability of the mortar

Figure 9. FTIR spectra of OPC mortar sample after 18 months of immersion in Na2So4

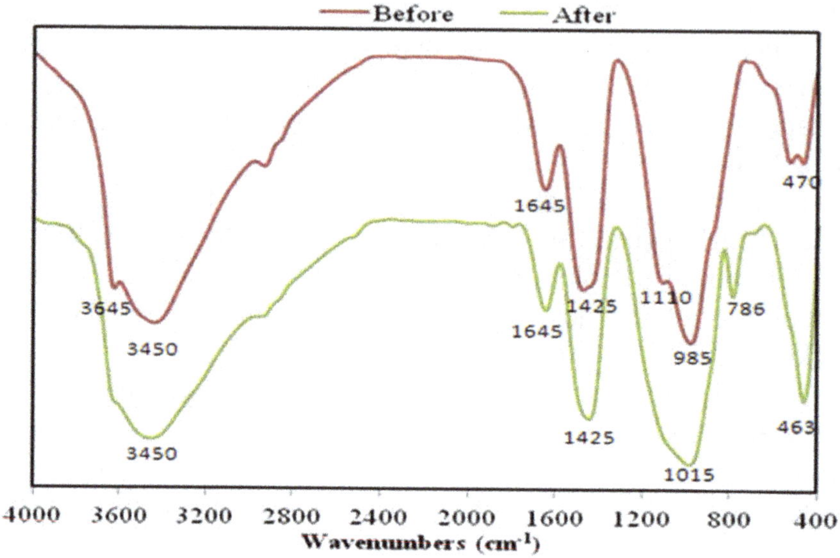

Figure 10. FTIR spectra of OPC mortar sample after 18 months of immersion in Na2So4

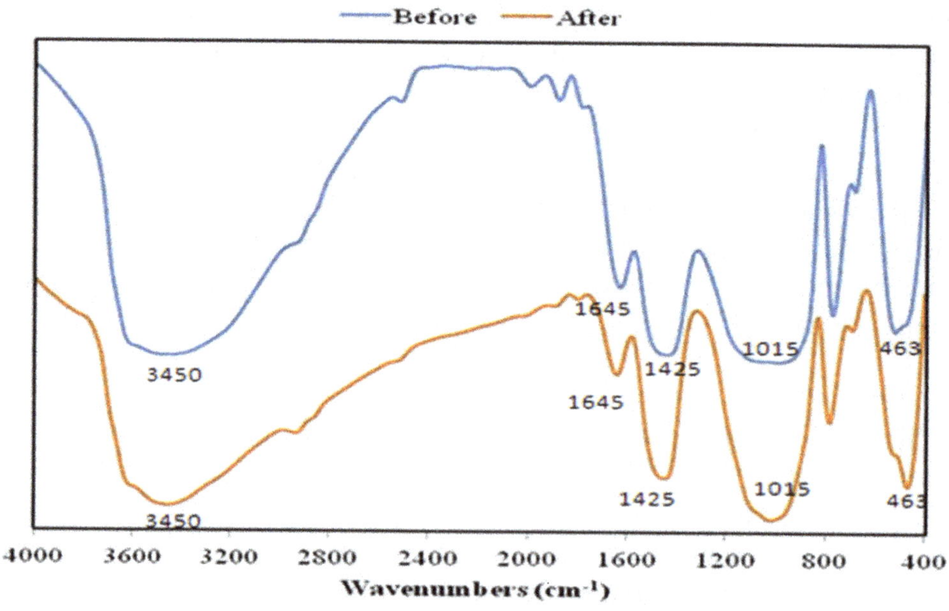

reduced and thus, the flow spread diameter of mortar also reduced. However, there is not much difference between the OPC mortar and ceramic mortar.

Furthermore, the study revealed that the ceramic mortar showed better resistance against aggressive environment than the OPC mortar. The performance of the ceramic mortar must have been influenced by the gradual pozzolanic reaction of ceramic powder and the fine particles filling the capillary pores that make the mortar denser, less permeable and more durable. Moreover, the low presence of CaO in the paste reduced the production of $Ca(OH)_2$ resulting in an increase in resistance against aggressive environments. The ceramic mortar demonstrated superior advantages with respect to visual appearance and mass change with low values of strength loss upon exposure to sulphate solution. Therefore, ceramic waste in the form of fine aggregates and fine powder can be used in mortar production with comparable strength and improvement in its fresh and hardened state properties compared with OPC mortar.

ACKNOWLEDGMENT

We, the editors greatly appreciate the exceptional contribution of Dr. Mostafa Samadi for carrying out this research and writing of the chapter. Indeed, members of his team, Dr. Nor Hasanah Abdul Shukor Lim, Dr. Ali Umara Shettima, Dr. Han Seung Lee, Dr. Ghasan Fahim Huseien and Dr. Nur Hafizah Abd Khalid deserve to be equally acknowledged for their co-authorship and immense collaboration in conducting/writing of the research.

REFERENCES

Ahmad, M. H., & Noor, N. M. (2011). Chemical attack of Malaysian pozzolans concrete. *Journal of Science and Technology, 1*, 11–24.

Bhutta, M. A. R., Farhayu Ariffin, N., Hussin, M. W., & Shukor Lim, N. H. A. (2013). Sulfate and sulfuric acid resistance of geopolymer mortars using waste blended ash. *Jurnal Teknologi, 61*(3), 1–5. doi:10.11113/jt.v61.1762

Binici, H., Shah, T., Aksogan, O., & Kaplan, H. (2008). Durability of concrete made with granite and marble as recycle aggregates. *Journal of Materials Processing Technology, 208*(1-3), 299–308. doi:10.1016/j.jmatprotec.2007.12.120

Budiea, A., Hussin, M. W., Muthusamy, K., & Ismail, M. E. (2010). Performance of High Strength POFA Concrete in Acidic Environment. *Concrete Research Letters*, *1*(1), 14–18.

Jaturapitakkul, C., Kiattikomol, K., Tangchirapat, W., & Saeting, T. (2007). Evaluation of the sulfate resistance of concrete containing palm oil fuel ash. *Construction & Building Materials*, *21*(7), 1399–1405. doi:10.1016/j.conbuildmat.2006.07.005

Khalil, H. P. S. A., Fazita, M. R. N., Bhat, A. H., Jawaid, M., & Fuad, N. A. N. (2010). Development and material properties of new hybrid plywood from oil palm biomass. *Materials & Design*, *31*(1), 417–424. doi:10.1016/j.matdes.2009.05.040

Mehta, P. K., & Monteiro, P. J. M. (1998). *Concrete micro structure, properties and materials*. New York: McGraw-Hill.

O'Farrell, M., Sabir, B. B., & Wild, S. (2006). Strength and chemical resistance of mortars containing brick manufacturing clays subjected to different treatments. *Cement and Concrete Composites*, *28*(9), 790–799. doi:10.1016/j.cemconcomp.2006.05.014

Toledo, R. D., Goncalves, J. P., Americano, B. B., & Fairbairn, E. M. R. (2007). Potential for use of crushed waste calcined-clay brick as a supplementary cementitious material in Brazil. *Cement and Concrete Research*, *37*(9), 1357–1365. doi:10.1016/j.cemconres.2007.06.005

Vejmelková, E., Keppert, M., Rovnaníková, P., Ondrác, M., & Robert, C. (2012). Properties of high performance concrete containing fine-ground ceramics as supplementary cementitious material. *Cement and Concrete Composites*, *34*(1), 55–61. doi:10.1016/j.cemconcomp.2011.09.018

Yunhong, C., Fei, H., Guang-lu, L., Longshuo, X., & Jianlong, H. (2014). Test research on effects of ceramic polishing powder on carbonation and sulphate-corrosion resistance of concrete. *Construction & Building Materials*, *55*, 440–446. doi:10.1016/j.conbuildmat.2014.01.023

Chapter 7
Effects of Palm Oil Fuel Ash as Micro-Filler on Interfacial Porosity of Polymer Concrete

ABSTRACT

The effects of palm oil fuel ash (POFA) as micro-filler on interfacial porosity and pore size distribution of polymer concrete (PC) is the main aim of this chapter. Beginning with a brief introduction about the topic, the materials and method used in this study are explained. Two categories of fillers were involved in this study, fine-micro filler (ground POFA and is paired with calcium carbonate), and coarse micro-filler (unground POFA and is paired with silica sand). It is revealed that the replacement of overall types of micro-filler at different filler content decreased the average pore diameter of PC significantly, except for PC incorporating unground POFA. Additionally, incorporation of fine-micro filler with dispersion characteristic could significantly reduce the interfacial porosity of PC as compared to incorporation of coarse micro-filler in PC.

INTRODUCTION

Polymer concrete (PC) has resins instead of cement as a concrete binder. The hardening of PC occurs without any water and polymerizes when added with additives, catalysts or accelerators. Additionally, since thermoset materials are hydrophobic materials, therefore, presence of water and high moisture content in aggregates has to be avoided especially during production of PC to enhance the compatibility and interaction between materials.

DOI: 10.4018/978-1-5225-8325-7.ch007

Generally, PC products become cost-effective especially when they contain micro-filler. Fillers normally come either from the natural sources (mineral filler) or are synthetically produced. However, the depletion of natural resources has become a concern, therefore, many researchers have shifted to use synthetic fillers, though this choice is more expensive. With the consideration for both environmental issues and sustainable development, utilization and modification of waste materials has become an interesting topic of research in modern production of polymeric materials. This had been initiated in the 1980s when fly ash was first studied and then further explored in the 1990s (Ohama, 2007). To date, the effectiveness of fly ash in enhancing the performance and durability of PC is proven, and is preferred for being cost-effective, toxic free and able to give good original fineness with low thermal coefficient. Moreover, it is readily available and is compatible with other materials in resin (Atzeni et al., 1990; Varughese and Chaturvedi, 1994; Gorninski et al., 2007).

This overview gives motivation to researchers to investigate the potentiality of Palm Oil Fuel Ash (POFA) usage in different types of concretes. Over the last few decades, the palm oil industry has grown up significantly in Southeast Asia such as Malaysia, Indonesia, and Thailand, and became an important agricultural-based industry. It is of no exemption to Africa where one of the major crops is also palm oil. This positive development was the reason for the researchers to utilize POFA in the cementitious system. However, POFA is a waste from agricultural plant and any agricultural based material has natural cellulose structure (Raveendran et al., 1996; Kaddami et al., 2006; Hafizah et al., 2014). This structure causes the PC to become hydrophilic, which also leads to very high and uneconomical resin consumption. Moreover, it deteriorates the PC quality even when fillers are used. Additionally, it becomes highly challenging in terms of reducing interfacial porosity in PC as well.

The objective of this study was to investigate the potentiality of POFA in reducing interfacial porosity in PC with low binder content using Mercury Intrusion Porosimetry (MIP) test. There existed several studies which were conducted by previous researchers that dealt with POFA in normal concrete. Therefore, two types of POFA were used in this study to compare their significant types in PC, i. e., ground and unground POFA. Additionally, ground POFA was paired with calcium carbonate as fine micro-filler, while, unground POFA was paired with silica sand as coarse micro-filler. All fillers were involved to investigate the function of filler on interfacial porosity. This work should eventually benefit researchers and fabricators in the field of PC production.

Materials and Method

Binder

The polymer binder used in this study was the Isophthalic polyester resin. It was chosen since it is the most economical and widely used in composites industry. The properties of the Isophthalic polyester resin are given in Table 1.

Aggregates and Fillers

The inert granular materials used were crushed coarse aggregates and natural river fine aggregates. The aggregates were oven-dried at 100 ± 5 °C since the polyester resin is hydrophobic. The final moisture content of the inert granular material was less than 0.1%. The coarse aggregates' size was from 10 to 12 mm.

The ground and unground POFA were designated as GPOFA and UPOFA, respectively. As mentioned earlier, GPOFA was paired with calcium carbonate ($CaCO_3$) and UPOFA with silica sand (Sand). The particle size distribution of all fillers is shown in Figure 1. The GPOFA was produced by grinding the POFA. Its further information can be found in Table 2.

Mix Proportions

Fifteen PC mixtures with POFA were prepared (Table 3) with the filler content fixed at 12% by weight of resin. The mixes were then combined with methyl ethyl ketone peroxide (MEKP) as the catalyst (hardener) and cobalt naphthenate (CoNp) as the initiator. The POFA, regardless of being unground

Table 1. Properties of isophthalic polyester resin

Properties	Isophthalic Polyester Resin
Density (g/m^3)[a]	1.1
Styrene monomer content (%)[a]	39-44
Viscosity (Brookfield, 25 °C, 60 rpm (cPs)[b]	538
Tensile strength (MPa)[b] ASTM D3039 (1995)	7
Tensile modulus (MPa)[b] ASTM D3039 (1995)	300

[a]Manufacturer data
[b]Viscosity and tensile properties were tested in this study

Figure 1. Particle size distribution for different types of fillers

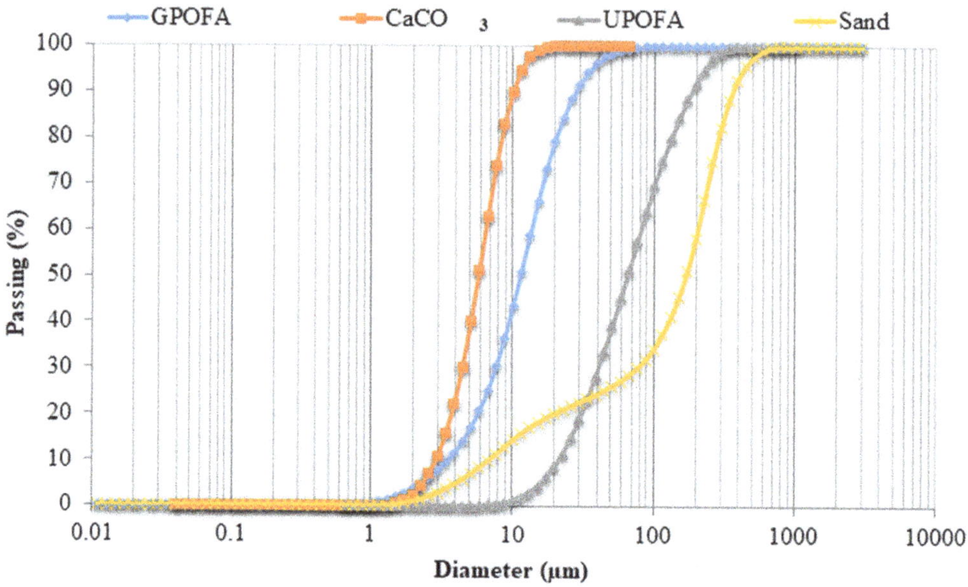

Table 2. Grinding mill information

Grinding Mill		
Dimension of grinding drum	:	711 mm (inside diameter) x 508 mm (inside length)
Number of rod bar	:	7 numbers
Dimension of rod bar	:	20 mm (diameter) x 480 mm (length)
Average diameter and length of rod bar	:	18 mm, 480 mm
Milling speed	:	30-33 rpm
Total milling rotational	:	4000 rotational per session
Capacity per session	:	5 kg/session

or grounded, took up 8 to 16% by weight of the PC spaces. Coarse aggregate content was fixed at 30% by weight. The design of PC mix proportion was carried out according to JIS A 1881 (2005). The nomenclature used was:

- **PC-GPOFA:** Polymer concrete incorporating ground POFA
- **PC-CaCO$_3$:** Polymer concrete incorporating calcium carbonate
- **PC-UPOFA:** Polymer concrete incorporating unground POFA
- **PC-Sand:** Polymer concrete incorporating silica sand

Table 3. Mix proportion of PC incorporating GPOFA and UPOFA

UP Binder		Filler						Fine Aggregates		Coarse Aggregates	
		POFA Filler		$CaCO_3$ Filler		Sand Filler					
%	kg/m³	%	kg/m³	%	kg/m³	%	kg/m³	%	kg/m³	%	kg/m³
12	134	8	78	8	216	8	198	50	1240	30	750
		10	97	10	270	10	248	48	1190		
		12	116	12	324	12	298	46	1141		
		14	136	14	378	14	347	44	1091		
		16	155	16	432	16	397	42	1042		

Note: polyester binder was mixed with inhibitor, 0.5% CoNp and 1% MEKP

Mercury Intrusion Porosimetry (MIP)

Mercury Intrusion Porosimetry (MIP) test was conducted to obtain porosity and pore size distribution using MIP equipment. Miniature of specimens (5 mm thick and the total weight were between 3.5 and 4.5 g) was subjected to low and high MIP pressure. The test demanded high precaution which should be taken since mercury is very harmful. Results of incremental pore volume, cumulative pore volume and average pore diameter were obtained at this stage.

RESULTS AND DISCUSSION

Figures 2 and 3 show the graphs of incremental pore volume versus pore distribution of PC in log normal term. Two obvious peaks have been observed at section I (pore diameter is small, ranging from 10,000 to 70,000 nm) and at section II (pore diameter is large, ranging from 70,000 to 120,000 nm). In general, the pore diameter is small. Also, results have shown that the finer micro-fillers performed better in reducing the PC pore volume. In addition, the incremental pore volume of PC-GPOFA and PC-$CaCO_3$ had reduced profoundly when the filler contents became higher (Figures 2a and 2b). Similar results were also observed for PC-Sand. However, the incremental pore volume decreased after adding more than 12% filler contents. This could be due to the progressive filling of binder particle gaps with sand. However, the overall decrease in pore volume was not obvious because the

Figure 2. Incremental pore volume versus normal-log pore diameter distribution for PC containing fine micro-filler (a)PC-GPOFA and (b) PC-CaCo3

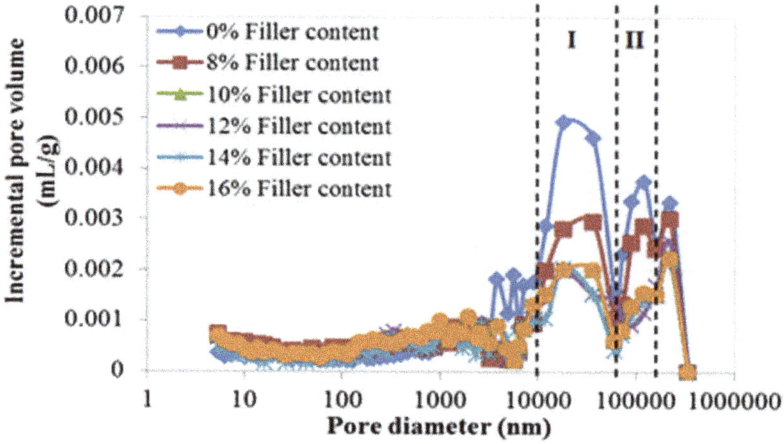

(a)

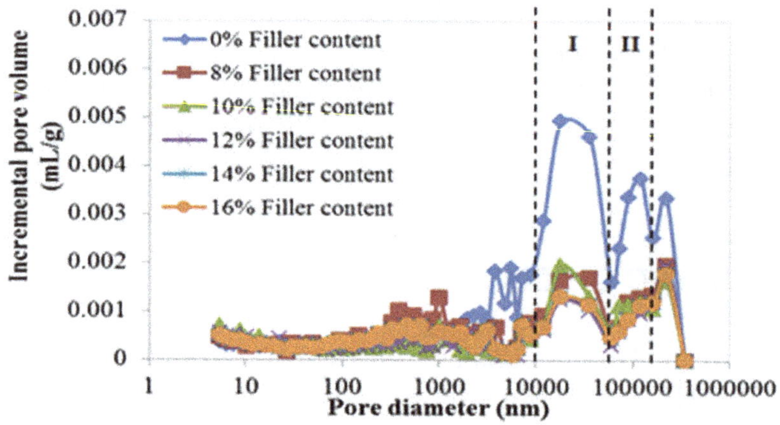

(b)

sand filler was occluded by the polymer resin. In PC-UPOFA, the reduction in pore volume was insignificant (Figures 3a and 3b) because it promotes the formation of large pores due to its natural open cellulose structure. Thus, GPOFA and calcium carbonate are preferred for their better micro-filler dispersing characteristics and ability to reduce the interfacial porosity.

Figure 3. Incremental pore volume versus normal-log pore diameter distribution for PC containing coarse micro-filler (a)PC-UPOFA and (b) PC-Sand

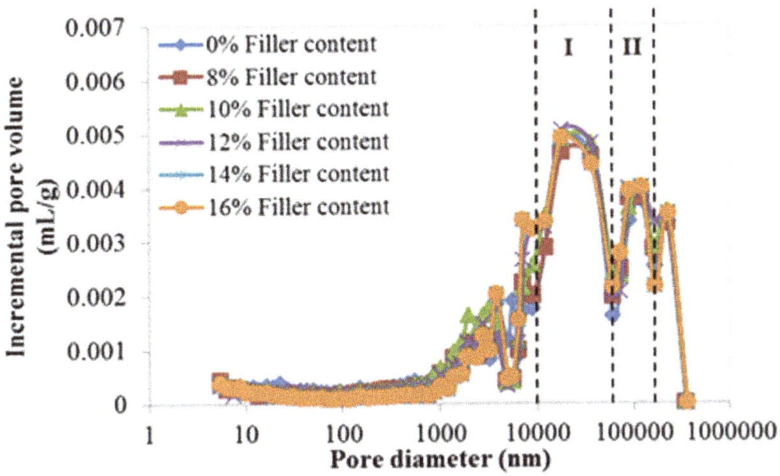

(a)

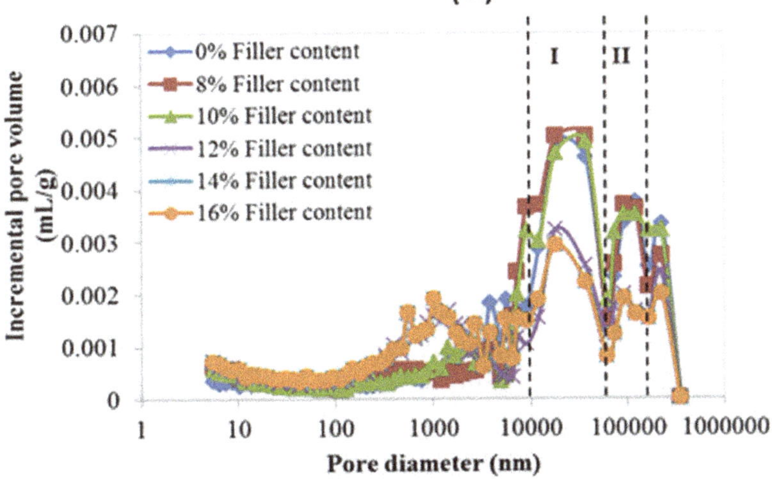

(b)

Figure 4 presents the cumulative pore volume-pore diameter behavior. The graph shows two main linear boundaries. The first or primary boundary is of greater interest in this case as significant changes of pores take place at the small pore diameter boundary. The second or secondary boundary is

Figure 4. Cumulative pore volume pore diameter behavior; primary and secondary boundary

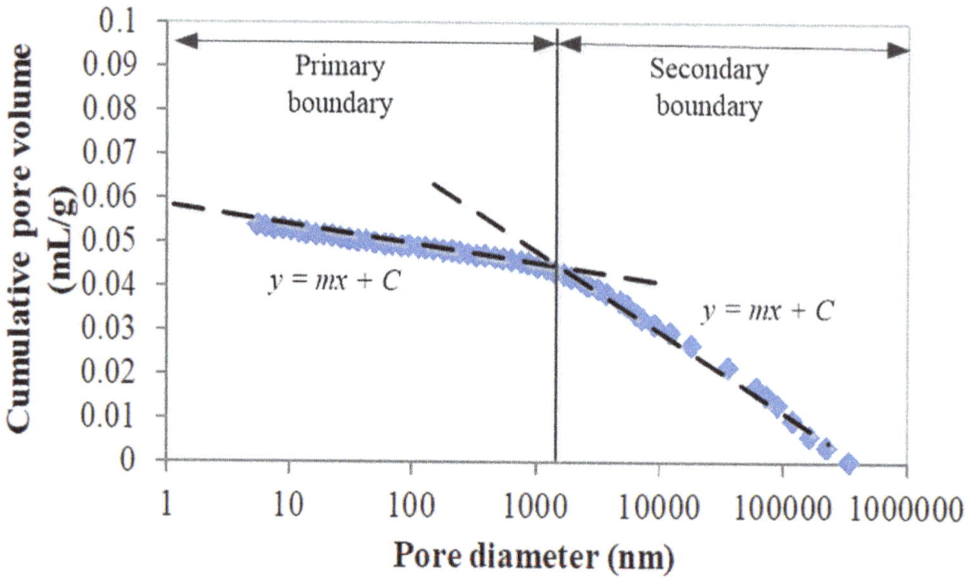

insignificant and therefore, ignored in this study. Figures 5 and 6 exhibit the cumulative pore volume-pore diameter behavior for all types of PC. From these Figures, it is evident that GPOFA and calcium carbonate have better quality as micro-filler than the rest because of the correlation between cumulative pore volumes and its diameter at primary boundary. The projected linear intersection, *C*, for PC incorporating fine micro-filler (GPOFA and calcium carbonate) decreased significantly when the filler contents increased due to their greater ability in reducing pore volume and diameter. This was based on two inherent characteristics, i. e., the fineness of the micro filler and the dispersing ability. Meanwhile, PC with coarse micro-filler (UPOFA and silica sand) did not show consistent behavior which is evident through its fluctuating projected linear intersection (*C* value). Nevertheless, the general patterns showed that both coarse and fine filler PCs had been prepared well in view of their consistent linear slope, *m*, and regression, R^2 as presented in Tables 4 and 5.

Figure 7 shows the effect of filler content on average pore diameter. This showed that the mean pore diameter had obviously reduced in higher filler content which could be due to an increased dispersing ability of micro-filler.

Figure 5. Cumulative pore volume pore diameter behavior for PC containing fine micro-filler (a) PCC-GPOFA and (b) PC-CaCO3

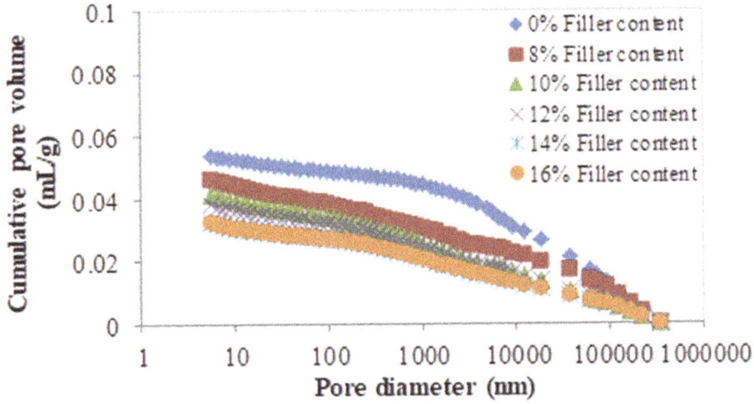

(a)

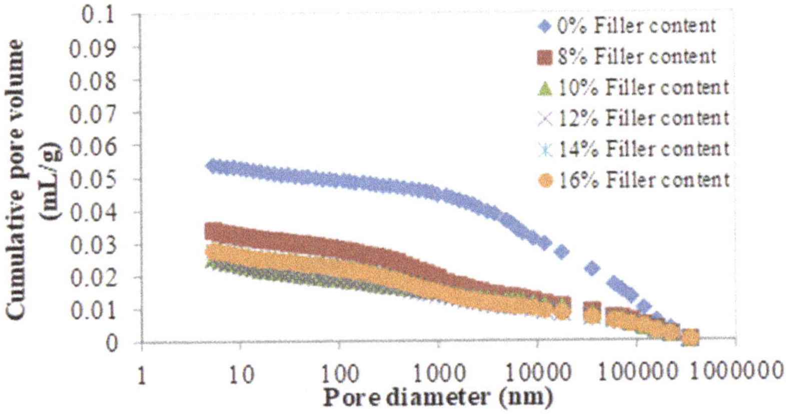

(b)

In fact, both GPOFA and calcium carbonate had dramatically reduced the average pore diameter. PC-sand also slowly reduced the pore volume when the sand was occluded by resin before it filled in the available gaps. PC-UPOFA had shown different patterns because of the poor filler option while PC-UPOFA showed no changes in pore diameter, which directly meant that the POFA filler had failed to function effectively as PC filler.

Figure 6. Cumulative pore volume pore diameter behavior for PC containing fine micro-filler (a)PC-UPOFA (b)PC-Sand

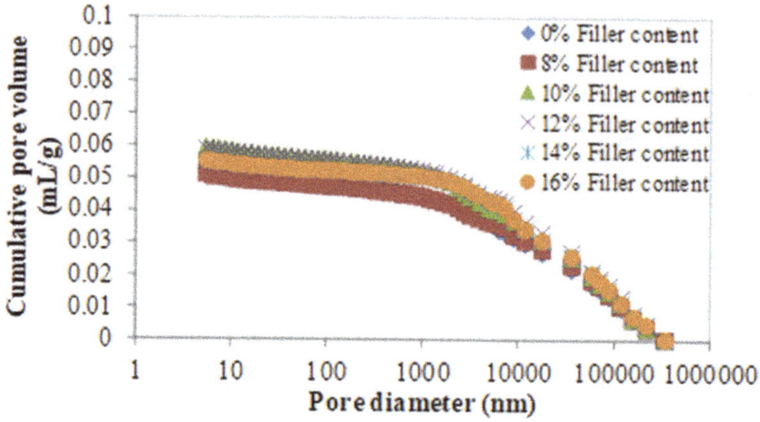

(a)

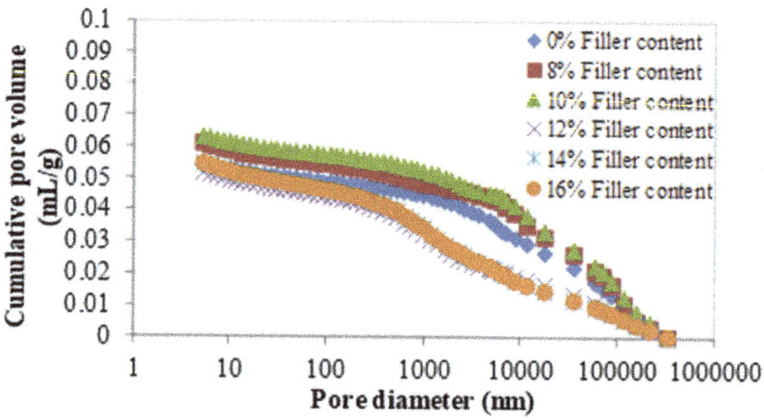

(b)

CONCLUSION

The fine micro-fillers in this study, i.e., GPOFA and calcium carbonate, were able to decrease PC interfacial porosity most effectively due to their fineness and dispersing ability. On the other hand, the coarse micro-fillers (sand and UPOFA) did not reduce the interfacial porosity significantly; the polymer resin occluded the sand filler before the particles could progressively fill into

Table 4. Correlation between cumulative pore volumes and its diameter at primary boundary for PC incorporating fine micro-filler

Filler Content (%)	PC-GPOFA			$PC\text{-}CaCO_3$		
	m	*C*	R^2	*m*	*C*	R^2
0	0.004	0.068	0.859	0.004	0.068	0.859
8	0.004	0.066	0.975	0.003	0.041	0.985
10	0.004	0.050	0.985	0.002	0.028	0.985
12	0.003	0.046	0.981	0.002	0.028	0.981
14	0.003	0.039	0.977	0.002	0.028	0.977
16	0.003	0.037	0.992	0.002	0.032	0.992

Table 5. Correlation between cumulative pore volumes and its diameter at primary boundary for PC incorporating coarse micro-filler

Filler Content (%)	PC-UPOFA			PC-Sand		
	m	*C*	R^2	*m*	*C*	R^2
0	0.004	0.068	0.959	0.004	0.068	0.859
8	0.004	0.065	0.917	0.005	0.075	0.838
10	0.005	0.076	0.926	0.005	0.079	0.833
12	0.004	0.075	0.891	0.005	0.064	0.971
14	0.004	0.071	0.855	0.005	0.068	0.971
16	0.004	0.071	0.855	0.005	0.068	0.971

Figure 7. Average pore diameter at different filler contents

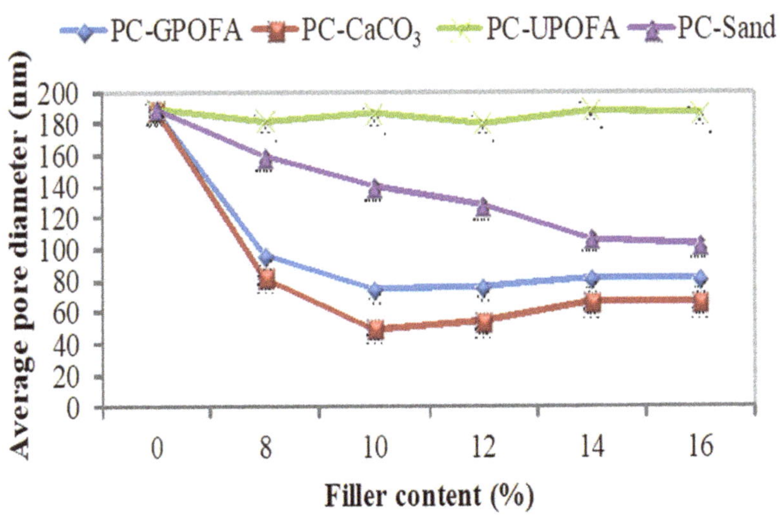

the positions in between the binder. Also, UPOFA promoted the formation of large pores due to its natural open cellulose structure and thus, had failed to reduce pore volume. To conclude, GPOFA can be potentially applied as micro-fillers for PC production.

ACKNOWLEDGMENT

We, the editors greatly appreciate the exceptional contribution of Dr. Nur Hafizah A. Khalid for carrying out this research and writing of the chapter. Indeed, members of her team, Dr. Nur Farhayu Ariffin, Dr. Nor Hasanah Abdul Shukor Lim, Dr. Mohammad Ismail, Dr. Ghasan Fahim Huseien, and Dr. Muhammad Ekhlasur Rahman deserve to be equally acknowledged for their co-authorship and immense collaboration in conducting/writing of the research.

REFERENCES

ASTM D3039. 1995. *Standard Test Method for Tensile Properties of Polymer Matrix Composite Materials*. West Conshohocken: ASTM International.

Atzeni, C., Massidda, L., & Sanna, U. (1990). Mechanical Properties of Epoxy Mortars with Fly Ash as Filler. *Cement and Concrete Composites*, *12*(1), 3–8. doi:10.1016/0958-9465(90)90030-2

Bignozzi, M., Saccani, A., & Sandrolini, F. (2000). New Polymer Mortars Containing Polymeric Wastes. Part 1. Microstructure and Mechanical Properties. *Composites. Part A, Applied Science and Manufacturing*, *33*(2), 205–211. doi:10.1016/S1359-835X(01)00093-8

Gorninski, J. P., Dal Molin, D. C., & Kazmierczak, C. S. (2007). Strength Degradation of Polymer Concrete in Acidic Environments. *Cement and Concrete Composites*, *29*(8), 637–645. doi:10.1016/j.cemconcomp.2007.04.001

Hafizah, N. A. K., Hussin, M. W., Jamaludin, M. Y., Bhutta, M. A. R., Ismail, M., & Azman, M. (2014). Tensile behaviour of Kenaf Fiber Reinforced Polymer Composites. *Jurnal Teknologi*, *69*(3), 11–15.

JIS A 1181. 2005. Test Methods for Polymer Concrete. Japan: Japanese Industrial Standard.

Kaddami, H., Dufresne, A., Khelifi, B., Bendahou, A., Taourirte, M., Raihane, M., & Sami, N. (2006). Short Palm Tree Fibers – Thermoset Matrices Composites. *Composites. Part A, Applied Science and Manufacturing*, *37*(9), 1413–1422. doi:10.1016/j.compositesa.2005.06.020

Ohama, Y. (2007). Recent Reserach and Development Trends of Concrete-Polymer Composites in Japan. In K.S. Yeon (Ed.). In *Proceedings of 12th International Congress on Polymer in Concrete*. Chuncheon, South Korea: Kangwon National University.

Raveendran, K., Ganesh, A., & Khilar, K. C. (1996). Pyrolysis Characteristics of Biomass and Biomass Components. *The Science and Technology of Fuel and Energy*, *75*, 987–998.

Varughese, K. T., & Chaturvedi, B. K. (1996). Fly ash as Fine Aggregate in Polyester based Polymer Concrete. *Cement and Concrete Composites*, *18*(2), 105–108. doi:10.1016/0958-9465(95)00006-2

Chapter 8
Evaluation of Electric Arc Furnace Oxidizing Slag Aggregates Quality and Development of Functional Concrete

ABSTRACT

Chapter 8 evaluates a most interesting and up-to-date topic of electric arc furnace oxidizing slag as aggregates' quality and development of functional concrete. A comprehensive introduction is given followed by a scientific method of stabilizing electric arc furnace (EAF). Oxidizing slag (EOS) is explained in brief. Subsequently, expansion mechanism of EOS and physical and chemical properties of EOS aggregates are covered in reasonable detail. A method for quantitative evaluation of free CaO contained in EOS that covers free CaO content as a function of aging period and open storage position for EOS and ERS samples are explained. Results indicated that the functional concrete using EOS aggregates satisfied the standards of slump, amount of air, its unit volume weight, and compressive strength. Moreover, x-ray irradiation experiment confirmed that the functional concrete using EOS aggregates showed a shielding performance approximately 20% higher than the typical concrete. These results verify that EOS has sufficiently good properties for use as concrete aggregate.

DOI: 10.4018/978-1-5225-8325-7.ch008

INTRODUCTION

Slag, which is a by-product inevitably generated during steel production, is recycled in an eco-friendly manner by the steel industry. Blast furnace slag is a slag which is generated during manufacturing of steel from magnetite in the blast furnace. Blast furnace slag is utilized as a concrete aggregate for embankments and roads (Faraone et al., 2009) as well as a substitute for raw materials in cement manufacture.

As the amount of slag generated annually increases, the possibility of recycling it as a high value product needs to be studied in order to utilize it efficiently in recycling industry and create economic benefits (Fernandez et al., 2007; Sheen et al., 2015; Sturm et al., 2009).

Blast furnace slag is actively developed and supplied as concrete aggregates and substitute for cement in concrete (Onoue et al., 2014; Motz et al., 2001). On the other hand, steel slag shows the problem of expansion and destruction during the hydration of cement in concrete due to unstable compounds present in it. This expansion and destruction during cement hydration occurs because of free CaO content present in it (Kuo and Shu, 2014; Wang, 2010). However, free CaO can be generated in steel slag during plasticity process. Free CaO content can vary according to temperature for plasticity in steelworks as well as cooling conditions. The melted slag is cooled slowly, aged in storage for approximately 3 months to a year to prevent expansion and destruction and is used as roadbed material.

In a study in Italy by Faleschini et al., 2016, some results of an experimental campaign about the potential use of EAF slag in cement-based materials were reported. The aim was to provide some insights about the important characteristics of the slag and its compatibility in concrete. Particularly, a mechanical characterisation was performed on compressive strength, tensile strength and elastic modulus showing the effects of substituting siliceous aggregates with slag. The performances were significantly enhanced in all cases. From that work, it was suggested that those improvements were gained both by a higher quality of the slag aggregates (in terms of density and strength), and by an improvement of the bond between EAF slag and the cementitious matrix. Similar results were also obtained recently by other authors, who obtained a singular morphology of the ITZ when EAF slag was used, which enhanced concrete mechanical properties.

The positive results obtained in terms of mechanical strength allowed designing concrete mixtures with the desired strength and workability leading to a significantly reduced environmental impact.

This process requires a wide storage yard and additional relevant costs. Electric arc furnace (EAF)-oxidizing slag (EOS) was studied. This slag was evaluated as an aggregate that does not pose problems pertaining to chemical expansion safety based on the Korean Industrial Standard (KS). However, the practical application of EOS aggregates was not implemented due to unstable property of material itself, for the product found in South Korea.

This study examined the physical and chemical properties of EOS and quantitatively evaluated the free CaO content in it and EAF-reducing slag (ERS), according to the period and location of aging using ethylene glycol. Moreover, the EOS aggregates were replaced depending on the EOS content to examine the physical properties of concrete and verify its quality characteristics as construction materials.

Method of Stabilizing Electric Arc Furnace (EAF)-Oxidizing Slag (EOS)

EOS should be sufficiently stabilized before it is used as an aggregate for roads or construction. The methods of stabilizing EOS are mainly classified into reforming method and aging method.

The reforming method generates other compounds that contain CaO where converter slag and EOS do not contain free CaO. Hence, no volume expansion occurs. In this method, additives (sand, waste foundry, red soil, etc.) and oxygen are added to the melted EOS to facilitate the slagging of free CaO and stabilize the slag. However, this stabilization method involves high cost because these additives should be added to molten slag to react with free CaO, while heat sources are also required to maintain the molten state when a large number of additives are used.

Although the reforming method is highly reliable in terms of stabilization, it was only studied and not commercialized because of problems such as the small amount of EOS processed, high processing cost, and prioritization for the production and quality of steel produced. This method can be implemented practically only if it is modified in such a way that additional heat sources are not required or minimized and cheap additives are used.

The aging method stabilizes EOS by slowly cooling high-temperature EOS and by crushing it according to the appropriate particle size. It is then left in the air before being used to artificially convert free CaO into $Ca(OH)_2$ and induce hydration in advance. This prevents expansion and destruction when slag is used. In this method, 3 m EOS is typically stored and left exposed to air. When the hydration and carbonation reactions between free CaO and water as well as with CO_2 are accelerated, the aging period is reduced. This aging period is reduced significantly in water or warm water compared to that in air. However, homogenization is unlikely to be performed during the aging process in water. In addition, although this method in warm water is implemented, it is not widely commercialized because it is not economically suitable.

The aging method in steam can make EOS more stable as an aggregate for roads compared to the aging method in air and facilitates the efficient use of limited land. At present, the entire steel industry in Japan uses aging methods in air or steam.

Expansion Mechanism of EOS

In terms of EOS, quicklime used as an adjunct to refine EAF materials such as pig iron and scrap iron, is frequently left behind in slag when they are insufficiently burnt. Figure 1 shows the compounds generated through the stabilization of quicklime in slag.

Existing studies (Wang, 2010) indicate that EOS is highly valuable for use as construction material because it contains large quantities of effective resources that can be recycled such as iron, carbon, and limestone. However, it was also reported that EOS contains a large quantity of free CaO, which is an unstable substance. Free CaO causes expansion and destruction because of hydration, thus leading to cracks in concrete during the hydration of concrete. When it reacts with water, its volume doubles and causes the expansion and destruction of slag.

Physical Properties of EOS Aggregates

The EOS produced in South Korea shows different physical properties depending on the region where it is produced. Table 1 describes the physical properties of EOS produced in Korea. It appears that these properties vary as slag is generated according to the types of pig iron output of steel manufacturers. Such differences are mainly based on density and absorption. In addition, the difference between EOS and ERS was verified.

Figure 1. Type of CaO compounds in the slag

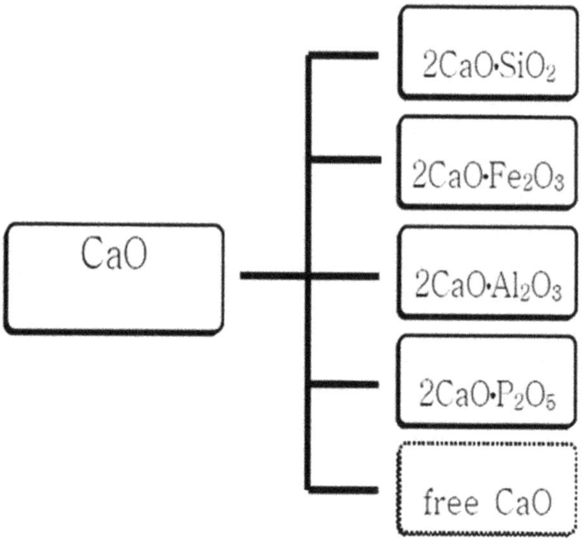

Table 1. Physical properties of the EOS in Korea

Type of Slag	A Steel Maker (Dangjin)			A Steel Maker (Incheon)		
	EOS		ERS	EOS		ERS
Aggregate Size	Fine	Coarse	Fine	Fine	Coarse	Fine
Density (ton/m3)	3.70	3.78	1.22	3.38	3.42	1.16
Fineness Modulus (%)	2.71	6.62	-	2.54	6.36	-
Absorption (%)	1.1	1.34	1.64	1.21	1.79	1.65

EOS: Electric arc furnace oxidizing slag, ERS: Electric arc furnace reducing slag.

Chemical Properties of EOS

Table 2 shows differences between the chemical compositions of EOS generated in Korea. It indicates that the chemical composition of EOS generated by the same steel manufacturer varies according to the region where it was produced. Moreover, a significant difference between EOS and ERS was observed. EOS contains approximately 30% Fe_2O_3 and 25% CaO, whereas ERS contains approximately 0.5% Fe_2O_3 and 60% CaO. It resulted from the different manufacturing processes between EOS and ERS.

Table 2. Chemical composition of the EOS in Korea

		Fe_2O_3	CaO	SiO_2	Al_2O_3	MnO	MgO	SO_3
A steel maker (Dangjin)	EOS	36.8	26.1	15.5	11.9	6.0	3.4	0.3
	ERS	0.5	58.4	20.3	12.7	1.5	5.0	1.6
A steel maker (Incheon)	EOS	34.3	20.7	21.9	10.4	8.2	4.3	0.2
	ERS	0.5	59.4	19.1	17.1	0.1	2.9	0.9

Existing studies report that the amount of free CaO (an expansion element) is insignificant in EOS. However, this study confirms that the physical and chemical properties of EOS vary according to the regions. In addition, it was determined that the amount of free CaO increases as its quantity increases in ERS.

Quantitative Evaluation of Free CaO Contained in EOS

The amount of free CaO present in EOS was quantitatively evaluated using ethylene glycol. The evaluation was conducted according to the aging periods and location of storing EOS.

According to the ethylene glycol method, the wet rendering should be controlled based on the composition of each material. This material is basically produced in unit ton, but for the purpose of experimental objective, 1g of material was sampled and the experiment was carried out. A 1 g sample was placed along with 50 mL of ethylene glycol in a 100 mL conical flask, which was then placed in a water bath maintained at 60 °C for 30 min. Each treated sample was filtered using two layers of No.5B filter bed through a Buchner funnel and washed thrice with 30 mL ethylene glycol.

The filtrate was then collected in an induction conical flask and titrated with N/10-HCl standard solution with 2 to 3 drops of Brome-cresol green solution added as an indicator. The terminal point was set when an N/10-HCl standard solution turned from blue to green. Using the amount of N/10-HCl standard solution consumed, the amount of free CaO can be calculated, as shown in Equation (1). Figure 2 shows the materials used for the experiments.

$$Free\,CaO(\%) = \frac{ml\,HCl \times normality\ of\ HCl}{10 \times sample\ weight} \times 28 \qquad (1)$$

where 28 = chemically combined water (Javellana and Jawed, 1982)

Figure 2. Materials used in the ethylene glycol method

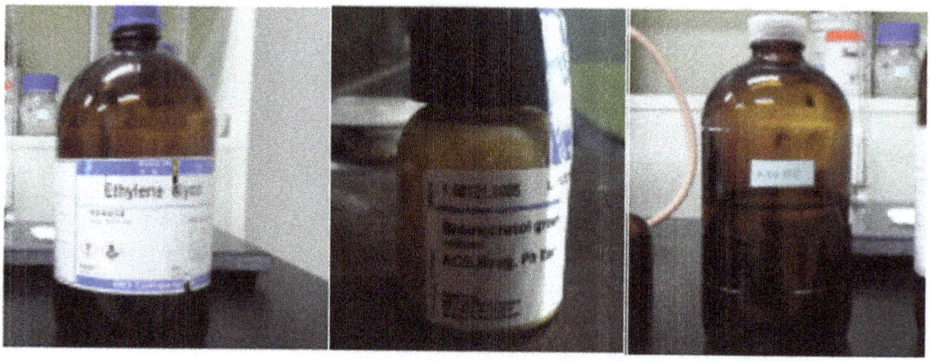

1. Ethylene glycol. 2. Brome-cresol green. 3. N/10-HCl.

Free CaO Content as a Function of Aging Period in the EOS and ERS Samples

Table 3 and Figure 3 show the results of these measurements. The free CaO contents in all the EOS samples were below 0.5%, and its content decreased as a function of aging period. On the other hand, while the ERS samples also showed remarkable decrease in free CaO contents as a function of aging period, it was always above 0.5%.

Free CaO Content as a Function of Open Storage Position for EOS and ERS Samples

Using the ethylene glycol method, CaO and free CaO contents in the EOS and ERS samples were quantitatively evaluated as a function of the open storage position (position within a depth of 1 m from the stacked slag). In

Table 3. Free CaO content at various aging periods in the EOS and ERS samples

Type of Slag	City	Sample	Free CaO Ratio (%)			
			1 Week	2 Week	3 Week	4 Week
EOS	Incheon	EOS 1	0.224	0.224	0.168	0.154
	Dangjin	EOS 2	0.491	0.421	0.421	0.421
ERS	Incheon	ERS 1	0.869	0.701	0.561	0.533
	Dangjin	ERS 2	1.795	1.753	1.738	1.682

Figure 3. Free CaO content at various aging periods in the EOS and ERS samples

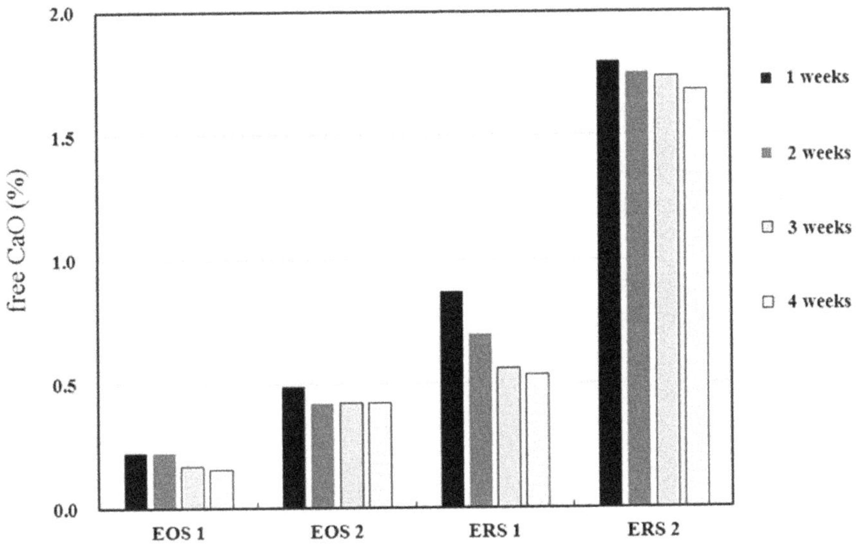

the case of electric furnace slag at A-Dangjin, an aggregate was pulverized for the experiments with a two day air-dried storage period. It was measured via XRF chemical analysis with the free CaO content (%) obtained by the ethylene glycol method. Although the CaO contents (%) in the EOS sample were identical at 26.1% at both storage positions, significant differences were observed in the free CaO contents for samples obtained from the two positions. The free CaO contents (%) of the EOS samples were 0.491% and 0.196%. In other words, both the values were below 0.5%. On the other hand, the free CaO contents in ERS samples at the two storage positions were also different (0.87%, and 1.74%). Table 4 shows the free CaO content and CaO content for various samples. The content was measured in both the inside (IN) and outside (OUT) samples. The reason is that when the sample is exposed to air, the free CaO content decreases due to reaction with moisture.

Table 4. CaO content and free CaO content of EOS and ERS by storage position

Type of Slag	Position	CaO (%)	Free CaO (%)
EOS	IN	26.1	0.491
	OUT	26.1	0.196
ERS	IN	60.1	0.869
	OUT	53.4	1.738

Properties of Functional Concrete Using EOS Aggregates

Experimental Plan

In an experiment examining the properties of functional concrete using EOS aggregates, the physical properties and radiation shielding performance were compared by using typical aggregates, magnetite, and EOS based on W/C 0.40%. Tests on the slump, amount of air, unit volume weight, and compressive strength were performed as part of the experiment on the physical properties of concrete. An X-ray shielding experiment was also conducted for radiation shielding performance test. Table 5 illustrates the mix details.

Materials Used

Ordinary Portland cement (OPC) was used in the experiment and its chemical properties are indicated in Table 6. EOS generated by steel manufacturer "A" located in Dangjin, washed river sand, and crushed granites were used as an aggregate, fine aggregate, and coarse aggregate, respectively. Table 7 presents the physical properties of aggregates. In addition, poly carboxylate ether was used as a super plasticizer.

Table 5. Mix design

Sample	W/C (%)	S/A (%)	Unit Weight (t/m³)								
			W	C	Fine Aggregates			Coarse Aggregates			Ad.
					NF	EOS	MA	NC	EOS	MA	
40N	40	47	165	413	810	-	-	911	-	-	3.3
40E					-	1257	-	-	1308	-	3.3
40M					-	-	1508-	-	-	1570	4.1
40EM					-	1257	-	-	-	1570	3.3

Table 6. Chemical composition of the cement

	CaO	SiO_2	Fe_2O_3	Al_2O_3	MgO	SO_3	Na_2O	K_2O
OPC	62	22	3	5	3	3	1	1

Table 7. Physical properties of the aggregates

	Normal Aggregates		EOS		Magnetite	
	Washed River Sand	Crushed Granites	Fine Aggregates	Coarse Aggregates	Fine Aggregates	Coarse Aggregates
Size (mm)	5	25	5	25	5	25
Density (t/m^3)	2.61	2.62	3.70	3.78	4.30	4.40
Fineness Modulus (%)	2.69	6.86	2.71	6.62	2.75	6.80
Absorption (%)	1.32	0.80	1.1	1.34	0.61	0.50

Experimental Method and Measurement Items

Concrete was classified into hardened and unhardened concrete in this experimental study. Tests on the slump, unit volume weight, and amount of air were conducted as part of the experiment on unhardened concrete, whereas those on compressive strength and X-ray irradiation were performed on hardened concrete. The compressive strength was tested by using specimens of $\varnothing 10 \times 20$ cm, which were obtained after standard curing at the age of 3, 7, and 28 days. The unit volume weight of each specimen was also measured.

Test Results on Physical Property of Concrete

Table 8 shows the results of experiments on unhardened concrete. The results of the slump test indicate that the concrete satisfies the slump standard of 150 ± 25 mm. Its unit volume weight is 3.14 t/m^3 in 40E with the highest value of 3.62 t/m^3 in 40M. This result verifies the difference according to the unit volume weight of aggregates. The results of the test on the amount of air show that it is the highest at 4.0% or greater at 40M.

Table 8. Test results of fresh concretes

	Slump (mm)	Air Content (%)	Unit Volume Weight (t/m^3)
40N	160	2.5	2.36
40E	160	3.1	3.14
40M	155	4.0	3.65
40EM	155	3.2	3.41

Figure 4 shows the result of tests on the compressive strength of hardened concrete. The compressive strength is the lowest (39.5 MPa) at 40N, 60.2 MPa at 40E, 55.5 MPa at 40M, and 61.5 MPa at 40EM. The compressive strength of EOS concrete is approximately 20 MPa higher than that of typical concrete.

X-Ray Irradiation Experiment

The fundamental equation for the X-ray irradiation experiment is Equation (2). Assume that the X-ray irradiating upon the shield has an intensity of I0, while that after the irradiation can be denoted as I. Here, x is the thickness of the shield, and μ_t is the effective attenuation coefficient, which represents the probability of the X-ray causing an arbitrary interaction with the material per unit length. Linear attenuation coefficient can be described using the Lambert law.

$$I = I_o \cdot e^{-\mu_t \cdot x} \tag{2}$$

where, I = dose rate at the surface of the shield facing toward the X-ray source

Figure 4. Test results of compressive strength

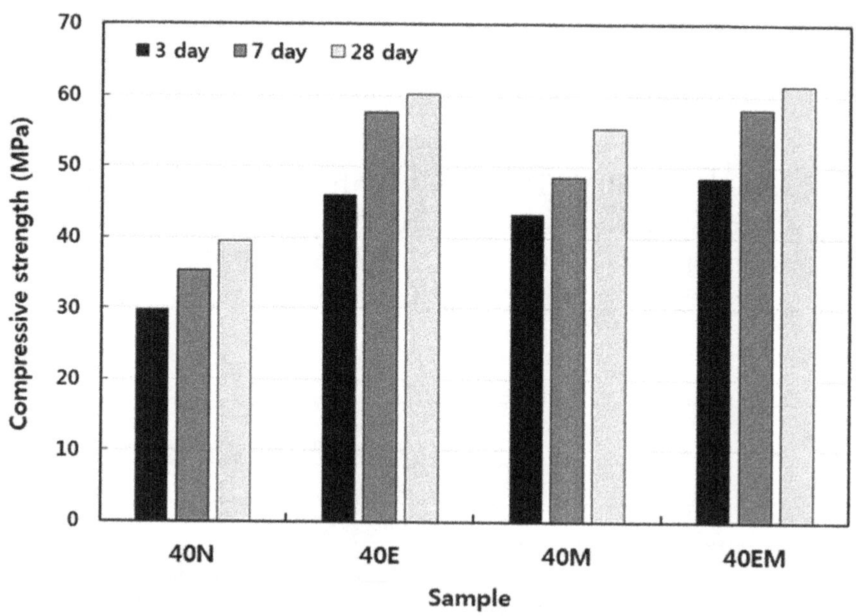

Table 9. Calculated data for the X-ray irradiation

Symbol of Specimen	Unit Volume Weight	Spectrum Data	IO/I	μt (cm-1)	Shielding Rate (%)
Source		1,712,234	1	-	-
40N	2.28	450,588	3.80	0.267	73.68
40E	3.21	148,117	11.56	0.490	91.35
40M	3.55	105,174	16.28	0.558	93.86
40EM	3.31	127,588	13.42	0.519	92.55

IO = dose rate at the surface of the shield opposite to the X-ray source
μ_t = effective attenuation coefficient, which is a constant representing the X-ray absorption ratio of the shielding material.
x = thickness of the shield

Table 9 shows the results from the X-ray shielding test. From the X-ray shielding ratio experiment, normal concrete specimens, 40N showed a shielding ratio of around 73%; EAF concretes, 40E showed a shielding ratio of around 91%; and magnetite concretes, 40M showed a shielding ratio of around 93%. If the unit volume weight of the concrete is high, the effective attenuation coefficient follows accordingly.

CONCLUSION

The evaluated quality-based characteristics of EOS aggregates indicates that the absorption ratio is within 2.0%, amount of CaO is below 40%, and basicity (CaO/SiO_2) is 2.0 or lower. The content of free-CaO in EOS is 0.5% or lower.

The results of a slump, air and unit volume weight and compressive strength on functional concrete using EOS aggregates indicate that it satisfies the standards of slump and amount of air and its unit volume weight as well as compressive strength. Moreover, the results of an experiment on the X-ray irradiation confirm that functional concrete using EOS aggregates shows a shielding performance approximately 20% higher than that of typical concrete and its performance is like that of magnetite concrete. These results verify that EOS has enough properties to be used as a concrete aggregate.

ACKNOWLEDGMENT

We, the editors greatly appreciate the exceptional contribution of Dr. Han Seung Lee for carrying out this research and writing of the chapter. Indeed, members of his team, Dr. Hee Seob Lim, Dr. Mohd Azreen Mohd Ariffin, Dr. Nor Hasanah Abdul Shukor Lim, and Dr. Mostafa Samadi deserve to be equally acknowledged for their co-authorship and immense collaboration in conducting/writing of the research.

REFERENCES

Faleschini, F., Brunelli, K., & Angelo Zanini, M. (2016). Electric arc furnace slag as coarse recycled aggregate for concrete production. *Journal of Sustain. Metall*, *2*(1), 44–50. doi:10.100740831-015-0029-1

Faraone, N., Tonello, G., & Maschio, S. (2009). Steelmaking slag as aggregate for mortars: Effects of particle dimension on compression strength. *Chemophere*, *77*(8), 1152–1156. doi:10.1016/j.chemosphere.2009.08.002 PMID:19740511

Fernandez Pereira, C., Luna Galiano, Y., Rodriguez Pinero, M. A., & Bale Parapar, J. (2007). Long and short-term performance of a stabilized/solidified electric arc furnace dust. *Journal of Hazardous Materials*, *148*(3), 701–707. doi:10.1016/j.jhazmat.2007.03.034 PMID:17459579

Javellana, M. P., & Jawed, I. (1982). Extraction of free lime in portland cement and clinker by ethylene glycol. *Cement and Concrete Research*, *12*(3), 399–403. doi:10.1016/0008-8846(82)90088-6

Kuo, W. T., & Shu, C. Y. (2014). Application of high-temperature rapid catalytic technology to forecast the volumetric stability behavior of containing steel slag mixtures. *Construction & Building Materials*, *50*, 463–470. doi:10.1016/j.conbuildmat.2013.09.030

Motz, H., & Geiseler, J. (2001). Products of steel slags an opportunity to save natural resources. *Waste Management (New York, N.Y.)*, *21*(3), 285–293. doi:10.1016/S0956-053X(00)00102-1 PMID:11280521

Onoue, K., Tokitsu, M., Ohtsu, M., & Bier, T. A. (2014). Fatigue characteristics of steel-making slag concrete under compression in submerged condition. *Construction & Building Materials*, *70*, 231–242. doi:10.1016/j.conbuildmat.2014.07.107

Sheen, Y. D., Lee, D. H., & Sun, T. H. (2015). Innovative usages of stainless steel slags in developing self-compacting concrete. *Construction & Building Materials*, *101*, 268–276. doi:10.1016/j.conbuildmat.2015.10.079

Sturm, T., Milacic, R., Murko, S., Vahcic, M., Mladenovic, A., Suput, J. S., & Scancar, J. (2009). The use of EAF dust in cement composites: Assessment of environmental impact. *Journal of Hazardous Materials*, *166*(1), 277–283. doi:10.1016/j.jhazmat.2008.11.015 PMID:19097693

Wang, G. (2010). Determination of the expansion force of coarse steel slag aggregate. *Construction & Building Materials*, *24*(10), 1961–1966. doi:10.1016/j.conbuildmat.2010.04.004

Chapter 9
Comparison Between Ordinary Portland Cement and Geopolymer Concretes Against Sulphuric Acid Attack

ABSTRACT

This chapter covers a comparison between ordinary Portland cement (OPC) and geopolymer concretes against sulphuric acid attack. An intensive introduction to the topic is given. Lack of study about high strength of self-compacting geopolymer concrete (SCGC) against sulphuric acid attack is also one of the problems. In this research, slag and ceramics were used as replacement of OPC. The aim was to study the durability of SCGC against sulphuric acid attack which mainly incorporated ground granulated blast-furnace slag (GGBFS) and ceramics waste as a binder. Methodology of the experimental program, with emphasis on preparation of materials and mix design is described. Testing procedure of GSCC is given. Durability test for sulphuric acid resistance and cost analysis are briefly explained. In conclusion, the sulphuric acid solution had no effect on the strength of concrete and the weight after being immersed in sulphuric acid solution for 28 and 42 days.

INTRODUCTION

Naik and Kumar (2010) reported that higher demand for concrete from the construction field has almost peaked in every country of the world including Malaysia. Therefore, developing countries are searching for new solutions and have been focusing more and more to recycle and reuse waste products to generate sustainable construction materials.

DOI: 10.4018/978-1-5225-8325-7.ch009

Ordinary Portland cement (OPC) production is considered a huge environmental problem because it requires large quantities of raw materials and fuel and contributes 8% of global CO_2 emissions (Scrivener and Olivier, 2012; Sanal, 2018). Furthermore, OPC is not a chemically stable material, as it deteriorates because of sulfate and sulfuric acid attacks as well due to elevated temperature exposure (Ariffin, 2013; Sanal et al., 2016). The use of pozzolans to replace part of OPC has become more attractive and important. Geopolymers (GP), the newly developed cement-free materials, are one alternative to overcome all the above-mentioned issues (Duxson, 2007).

Previous research study has shown that OPC concrete has a very low durability against sulphuric acid attack compared to geopolymer concrete (Shankar, 2012). Durability refers to the ability of concrete to resist deterioration from the environment or service to which it is exposed. Suitably proportioned concrete that is properly placed, finished, and cured can endure without significant distress throughout its service life (Taylor, 2013). This is because the penetration of sulfate ions in solution causes change in the chemical composition and microstructure of the concrete. These changes may vary in type or severity, but commonly include extensive cracking, expansion and loss of bond between the cement paste and the aggregates. Alteration of paste composition results in monosulfate phase which converts into ettringite and at later stages forms gypsum. The required additional calcium is provided by calcium hydroxide and calcium silicate hydrate present in the cement paste. The effect of these changes leads to an overall loss of concrete strength (Ferraris, 2006).

In addition, geopolymer binders have been reported as being acid resistant and thus are promising alternatives for sewer pipe manufacture compared to OPC concrete (Song, 2005). A lack of needed literature study about high strength self-compacting geopolymer concrete against sulphuric acid attack is also one of the problems.

In this research, slag and ceramics were used as replacement to OPC. Slag contained high calcium, which is suitable for aggressive environment (Islam, 2010), while the ceramic containing high silica can make the concrete more durable with more binder gel (Paratibha, 2015). There is also pressure to provide a better quality concrete to the construction sector which preserves natural resources and air quality. It is expected that this self-compacting geopolymer concrete (SCGC) can solve all the issues described in OPC (Srinivas, 2015).

The aim of this research was to study the durability of SCGC against sulphuric acid attack. SCGC mainly incorporated Ground Granulated Blast-Furnace Slag (GGBFS) and ceramics wastes as a binder. These parameters include determining the optimum mix design for geopolymer concrete containing GGBFS and ceramics as binders, their durability performance against sulphuric acid attack and the effect of geopolymer concrete after immersion in sulphuric acid solution.

Cube specimens with 100 mm x100 mm x100 mm dimensions were prepared and cured at ambient temperature. Compressive strength test was conducted to study the engineering properties of geopolymer concrete. The cubes were tested for compressive strength after being immersed in sulphuric acid solution at ages of 28 days and 42 days. The concentration of sulphuric acid solution has a pH of 3.0.

METHODOLOGY

Preparation of Materials

Materials used in this research study basically consisted of GGBFS, ceramics, fine (Sand) and coarse aggregates (Gravel 10 mm) and alkaline solution (Sodium Hydroxide and Sodium Silicate). The grading of fine and coarse aggregates followed the ASTM standard C33 (Standard Specification for Concrete Aggregates).

The GGBFS was collected from slag factories in southern parts of Malaysia while the ceramic waste was collected from a ceramic factory in Johor. After collection, the ceramic was crushed in a crusher jaw to make it finer and easier to sieve. The crushed ceramic waste was sieved by a mechanical sieve shaker for a sufficient period. It was then grounded in a Los Angeles Abrasion Machine for at least 8 hours (for each 4 kg) using a steel bar (18 mm diameter and 800 mm long). Only the finest ceramic was used in the mixing which was kept tightly secured in a container.

The coarse aggregates (gravel) were collected from the Structure and Materials Laboratory at Universiti Teknologi Malaysia (UTM), Malaysia. The gravel was sieved through a mechanical sieve shaker and only the gravel that passed through 10mm sieve were collected and used in the mixing.

Mix Design

Cube Samples

The mix proportions of geopolymer self-compacting concrete and the control OPC concrete samples are shown in Tables 1 and 2, respectively.

Testing Procedure of GSCC

In this research, the slump flow and L-Box test were performed to ascertain the self-compacting capabilities. All those tests were conducted in accordance with **E**uropean **F**ederation of **N**ational **A**ssociations **r**epresenting producers and applicators of specialist building products for **c**oncrete (EFNARC) guidelines. The slump flow test was used to assess the horizontal free flow of SCC in the absence of obstruction. On lifting the slump cone filled with concrete, the concrete begins to flow. The time T_{50cm} is a secondary indication of flow. It measures the time taken (in seconds) from the time when the cone is lifted to the instant when horizontal flow reaches a diameter of 500 mm.

The passing ability is determined by the L-box test as shown in Figure 2. The vertical section of the L-box is filled with concrete, and then the gate lifted to let the concrete flow into the horizontal section. The height of concrete at the end of the horizontal section is expressed as a proportion of that remaining in the vertical section (H_2/H_1). This is an indication of passing ability.

Table 1. Mix proportion of GSCC

	Slag (kg/m³)	Ceramic (kg/m³)	F.A (kg/m³)	C.A (kg/m³)	SP (kg/m³)	Na_2SiO_3 (kg/m³)	NaOH (kg/m³)	Extra Water (kg/m³)
GSCC	600	600	860	1.11	86.7	393.3	156.7	136.7

Table 2. Mix proportion of OPC concrete

	W/B Ratio	Water (kg/m³)	OPC(kg/m³)	FA(kg/m³)	CA(kg/m³)
OPC (100%)	0.59	192	384	898	861

Figure 1. Procedure for slump-flow test

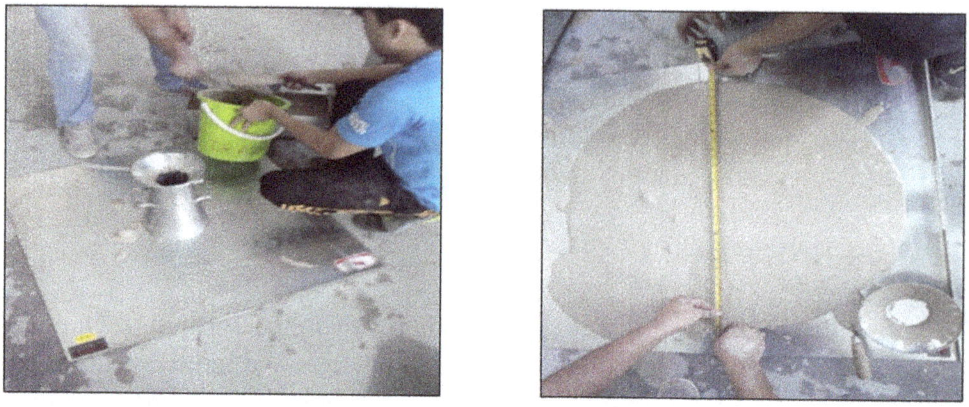

Figure 2. Procedure for l-box test

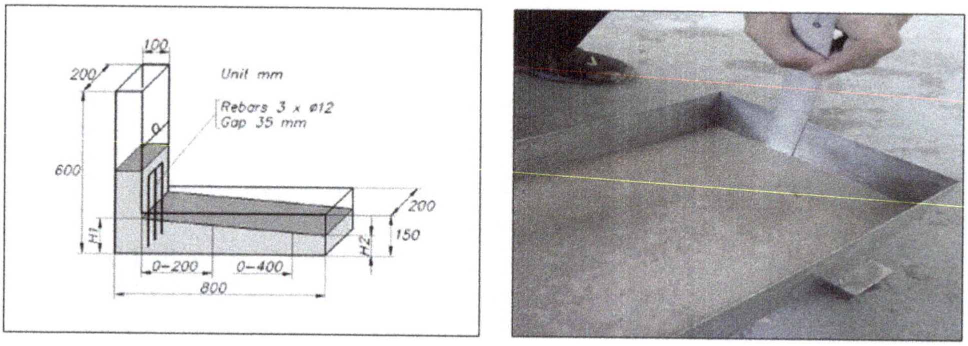

Durability Test on OPC Concrete and GSCC

It is well known that concrete is not fully resistant to acids. All acids will have an impact on the performance of concrete. The rate of speed of action may differ, but they eventually disintegrate the concrete. Six cube specimens of size 100 mm x100 mm x100 mm were prepared and immersed in sulphuric acid solution (pH 3) for 28 and 42 days.

Figure 3 shows the compressive strength comparison between OPC concrete and geopolymer concrete (GSCC) after acid immersion for 28 days. GSCC showed better strength compared to OPC concrete at ages of 28 and 42 days. The high percentage of calcium in the geopolymer binder (slag) produced better concrete in terms of durability and strength. The composition

Figure 3. Compressive strength against days of immersion

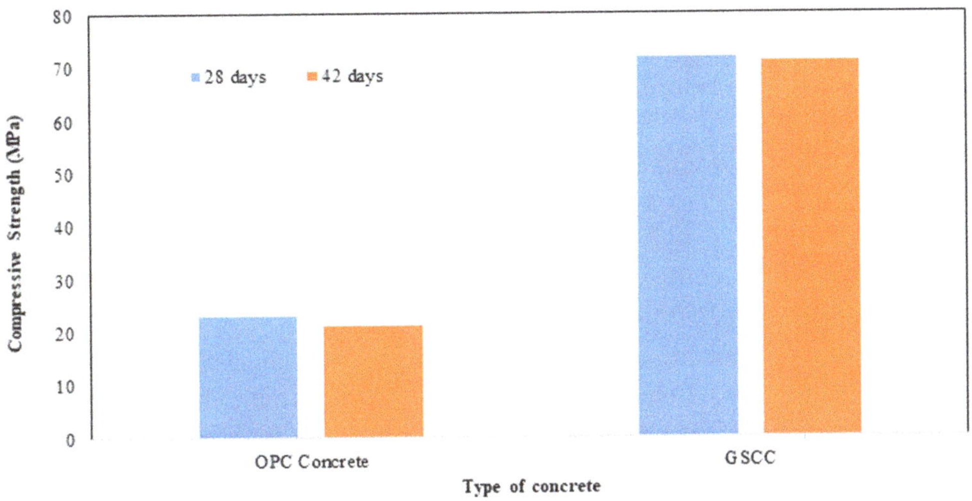

and microstructure of the OPC concrete changed because of the penetration of sulfate ions present in the sulphuric acid solution, thus forming gypsum. These changes result in an overall loss of concrete strength.

Standards

All the standards that relate to any testing for mechanical properties and microstructure characterization of GSCC follow all the same relevant standards that are applied to the OPC concrete.

Cost Analysis

The cost analysis calculations are based on geopolymer concrete using waste materials as a binder replacing 100% OPC. The cost of geopolymer concrete is kept RM350.00 per m^3 which does not consider the cost of alkaline solution's amount in bulk volume. If we can apply this geopolymer concrete product to the construction industry, the price of the alkaline solution can be reduced further.

Besides that, the cement-free geopolymer concrete carries a high early strength, durability and resistance against attacks from aggressive environments compared to OPC concrete.

CONCLUSION

The following conclusions can be drawn about geopolymer self-compacting concrete against sulphuric acid attack from this study:

1. The most optimum mix design for granulated blast furnace slag and ceramic powder as a cement replacement is 50% slag and 50% ceramic powder.
2. 8 M Sodium Hydroxide (NaOH) solutions gives better workability as compared to 12M NaOH solution.
3. Samples containing slag has low workability because of its low relative density which causes an increase in paste volume. Thus, more water is needed for the self-compacting concrete, which can lead to low compressive strength.
4. The sulphuric acid solution had no effect on the strength of concrete and its weight after being immersed in sulphuric acid solution (pH = 3) at ages of 28 and 42 days.

ACKNOWLEDGMENT

We, the editors greatly appreciate the exceptional contribution of Dr. Mohd Azreen Mohd Ariffin for carrying out this research and writing of the chapter. Indeed, members of his research team, Mr. Shafiq Ishak, Dr. Nur Farhayu Ariffin, Dr. Nor Hasanah Abdul Shukor Lim, Dr. Ali Umara Shettima and Dr. Kharunisa Muthusamy deserve to be equally acknowledged for their co-authorship and immense collaboration in conducting/writing of the research.

REFERENCES

Aggarwal, P., Singh, R. P., & Aggarwal, Y. (2015). Use of nano-silica in cement based material. *Civil and Environmental Engineering*, *2*, 1–11.

Ariffin, M. A. M., Bhutta, M. A. R., Hussin, M. W., Mohd Tahir, M., & Aziah, N. (2013). Sulfuric acid resistance of blended ash geopolymer concrete. *Construction & Building Materials*, *43*, 80–86. doi:10.1016/j.conbuildmat.2013.01.018

Duxson, P., Provis, J. L., Lukey, G. C., & van Deventer, J. S. J. (2007). The role of inorganic polymer technology in the development of 'green concrete'. *Cement and Concrete Research*, *37*(12), 1590–1597. doi:10.1016/j.cemconres.2007.08.018

Ferraris, C. F., Stutzman, P. E., & Snyder, K. A. (2006). *Sulfate Resistance of Concrete: A New Approach, R&D Serial No. 2486*. Skokie, IL: Portland Cement Association.

Hussin, M. W., Bhutta, M. A. R., Azreen, M., Ramadhansyah, P. J., & Mirza, J. (2015). Performance of blended ash geopolymer concrete at elevated temperatures. *Materials and Structures*, *48*(3), 709–720. doi:10.161711527-014-0251-5

Islam, M. M., Islam, M. S., Bipul Chandra, M., & Rafiqul Islam, M. (2010). Strength behavior of concrete using slag with cement in sea water environment. *Journal of Civil Engineering*, *38*(2), 129–140.

Naik, T. R., & Kumar, R. (2010). Geopolymer concrete for sustainable developments: opportunities, limitations and future needs. *Proceedings of Third International Conference on Sustainable Construction Materials and Technologies*, 1-8.

Olivier, J. G. J., Janssens-Maenhout, G., & Peters, J. A. H. W. (2012). *Trends in global CO_2 emissions*. The Hague: PBL Netherlands Environmental Assessment Agency.

Peter Taylor, P. T. (2013). *Durability of Concrete*. Transportation Research Board of the National Academies.

Sanal, I. (2018). Performance of macro-synthetic and steel-fiber reinforced concretes with emphasis on mineral admixture addition. *Journal of Materials in Civil Engineering*, *30*(6), DOI: MT.1943-5533.0002292 doi:10.1061/(ASCE)

Sanal, I., Ozyurt Zihnioglu, N., & Hosseini, A. (2016). Characterization of hardened state behavior of self-compacting fiber-reinforced cementitious composites (sc-frcc's) with different beam sizes and fiber types. *Journal of Composites Part B: Engineering*, *105*, 35–45.

Sanni, S. H., & Khadiranaikar, R. B. (2012). Performance of geopolymer concrete under severe environmental conditions. *International Journal of Civil and Structural Engineering*, *3*(2), 396–407.

Scrivener, K. L., & Kirkpatrick, R. J. (2008). Innovation in use and research on cementitious material. *Cement and Concrete Research, 38*(2), 128–136. doi:10.1016/j.cemconres.2007.09.025

Song, X. J., Marosszeky, M., Brungs, M., & Munn, R. (2005). Durability of fly ash based Geopolymer concrete against sulphuric acid attack. *Proceedings of 10DBMC International Conference on Durability of Building Materials and Components*, TT3-48

Vasam, S., Jagannadha Rao, S. K., & Seshagiri Rao, M. V. (2015). Self-compacting concrete (SCC) as green concrete with recycled concrete aggregates (RCA) –an experimental study. *International Journal of Research in Engineering and Technology, 4*(13), 1–7.

Chapter 10
Self-Healing Mortar

ABSTRACT

The cost of repairing cracked concrete is expensive as it requires special repair materials and skilled labour. Thus, the developments of new materials, like self-healing materials, are highly needed to repair cracks automatically and to restore or even increase concretes' strength to prolong its service life. The aim of this chapter was to investigate the performance of epoxy resin without hardener as a self-healing agent in mortar. A detailed introduction of self-healing mortar is given followed by a problem statement. The epoxy resin as a self-healing material is also explained briefly. Self-healing concept is also discussed in detail followed by the experimental program. Results revealed that the epoxy resin without hardener as a healing agent performed effectively as the compressive strength and ultrasonic pulse velocity of 365 days old cracked mortar samples regained the initial reading with prolonged curing time.

INTRODUCTION

Concrete is a strong, relatively cheap construction material and is presently the most widely used material in the construction industry. The main constituent that contributes to concrete strength is Portland cement. It is estimated that cement (Portland clinker) production alone contributes about 7% of the global carbon dioxide emissions due to the burning of limestone and clay at a temperature of ~1500°C as well as fossil fuel. During this process, calcium

DOI: 10.4018/978-1-5225-8325-7.ch010

carbonate ($CaCO_3$) is converted to calcium oxide (CaO) and carbon dioxide is released (Worrell et al., 2001). From an environmental standpoint, concrete does not appear to be a sustainable material (Gerilla et al., 2007). Moreover, improperly manufactured concrete may experience a shorter service life as it can easily develop cracks due to excessive loading applications and other environmental causes. Hence, a good quality of concrete is needed which not only prolongs its service life but would also reduce the production of cement.

The major problem faced by concrete structures is cracking. It can cause severe damage such as corrosion of steel reinforcement that can lead to the deterioration of structures. A number of cracks with different morphology and size could appear during construction and the design life of structure (Zhang et al., 2011). Small cracks or micro-cracks need to be repaired before they become a major or macro-crack. There are many types of repair materials and methods that can be selected and used to repair these micro-cracks. However, in certain cases, micro-cracks will still exist in the concrete structure and if they cannot be repaired effectively, the performance of the structure is compromised and its service life will be shortened. The cost of repairing concrete is expensive as it requires raw materials such as cement and skilled labours.

In many concrete structures, tensile forces can cause cracks and these can occur relatively soon after the structure is built. Repairing cracks in conventional concrete structures usually involves applying mortar which is bonded to the damaged surface. Sometimes, the mortar needs to be installed into the existing structure to ensure it does not fall off. The bonding property between host concrete and new repair materials needs to be carefully verified to avoid another problem leading to extra repair cost. Repairs can be particularly time consuming and expensive; often it is very difficult to gain access to the structure to make repairs, especially if they are underground or at a great height. Therefore, the development of new technologies and materials is essentially needed that can automatically repair cracks by giving longevity to the structures. Such materials will eventually reduce maintenance costs and fulfil the need of sustainability.

The self-healing concrete had been investigated since 1996 by Stefan Jacobsen and Erik J. Sellevold from Norway. In their research paper, they investigated the deterioration of self-healing concretes caused by internal cracking. Besides, Nynke ter Heide (2005) from Delft University of Technology had discussed self-healing process of concrete in detail. This led to more investigations conducted by other researchers in the area of self-healing concrete. The occurrence of autogenous healing or self-healing of cracks in concrete was recognized in a previous study by Reinhardt and Jooss (2003). On the other hand, the research of Jonkers and Schlangen (2008) had developed a microorganism that can heal the cracks automatically.

Self-healing concrete refers to the ability of concrete to automatically repair the cracks without any external intervention. In concrete, the most common self-healing agent is bacteria. The bacteria used in some studies are known as microorganisms that can survive in alkaline environment of concrete (pH 13 to 14), provided the air and enzyme are available. However, in this study, a different type of self-healing agent was used to incorporate with the mortar mix which eventually should heal the cracks.

Problem Statement

In most of the studies, an ureolytic bacterium of the genus Bacillus was used as a healing agent for the biological production of calcium carbonate. The mechanism of self-healing by using bacteria is the formation of calcium carbonate from the bacteria, based on the enzymatic hydrolysis of urea to ammonia and carbon dioxide (Hager et al., 2010). This reaction causes pH to increase from neutral to alkaline conditions resulting in the formation of bicarbonate and carbonate ions which precipitate in the presence of calcium ions to form calcium carbonate minerals (Jonkers et al., 2010).

Bacteria are good self-healing agents, but the acceptance of using it in concrete has been a big issue among contractors. The perceptions of using bacteria in the concrete become a serious concern perhaps because bacteria itself have a negative effect on human health. Moreover, to find and culture the bacteria is not an easy task as several conditions need to be considered for bacteria to survive. On top of the list is the availability of bacteria once the concrete is hardened. It becomes a main concern as bacteria are a sensitive microorganism. During the process of concrete mixing, the applied pressure can somehow destroy the bacteria. The alternative self-healing agent that shows promising results in healing the crack automatically is epoxy resin. This study focused on the self-healing performance of epoxy resin as a self-healing agent.

Thermoset Polymer: Diglycidyl Ether of Bisphenol – A Type of Epoxy Resin

Epoxy resin is a family of thermosetting polymer materials which do not give off reaction products when they cure and have low shrinkage. Moreover, it has good adhesion to other materials, good chemical and environmental resistance, good chemical and good insulating properties (Ariffin, 2016). In the present study, the high viscosity of epoxy resin namely Diglycidyl Ether of Bisphenol A-type was used. Epoxy resin has been used as one of the repair materials to repair concrete for many decades. The use of epoxy resin which is an external repair process entails the injection of epoxy resin solution together with a hardener into existing cracks in concrete structure. This method is well known and usually used for concrete structure repairs suffering from cracks.

Epoxy resin with different level of viscosity can repair cracks with different widths. Upon injection, a suitable pressure is applied which leads the epoxy resin to flow through the cracks. Besides using epoxy resin as a repair material, it can also be used as an additive or replacement material in concrete mixtures. The addition of epoxy resin in the concrete/mortar is categorized as polymer-modified concrete/mortar. In this study, the epoxy resin was used in mortar specimens and denoted as epoxy-modified mortar. Usually, the use of epoxy resin with the hardener has been common practice. The hardener functions as a hardening agent for the epoxy resin which strengthens the binders in concrete/mortar.

However, in this study, the epoxy resin was used without hardener as a self-healing agent. The understanding was that epoxy resin can harden inside the concrete/mortar by reacting with hydroxyl ions. Hydroxyl ions are produced from cement hydration and react with epoxy resin to heal the cracks. The rationale for not using hardener was to let the epoxy resin stay in the same state for later reaction when cracks occurred. The unhardened epoxy resin would then react with the hydroxyl ions to repair the cracks automatically without any external intervention.

Self-Healing

Self-healing ability of the material to heal, recover and repair the cracks automatically without any internal intervention is seen as an interesting occurrence. The additional materials such as bacteria and epoxy resin have a unique function to react and produce the element that can heal the cracks whenever they occur inside the concrete structure. Even when the crack size and width are limited to the micro-crack with a diameter between 0.1 mm and 0.3 mm, it has happened (Palin et al., 2015; Edvardsen, 1999).

The performance of epoxy resin as a self-healing agent was investigated by non-destructive test and compressive strength test as shown in Figure 1. The non-destructive test was selected to determine the self-healing effect by checking the formation of cracks produced at three different curing times. This technique has been previously used by Zhong and Yao (2008) for the purpose of self-healing evaluation. The first crack was induced at 28 days, which was immediately after the maturity of mortar. The second and third periods of the cracks production were at ages of 180 days and 365 days of the curing period, respectively. These durations were chosen in order to see the effectiveness of the epoxy resin as a healing agent after a certain period. The main purpose of this test method was to simulate the real situation in construction where sometimes the cracks occur at different and unexpected times.

There are two types of loads applied to the mortars; 50% and 80% of the ultimate load. The mortar that preloaded at 50% of ultimate load, was denoted

Figure 1. Procedure of producing crack in mortar

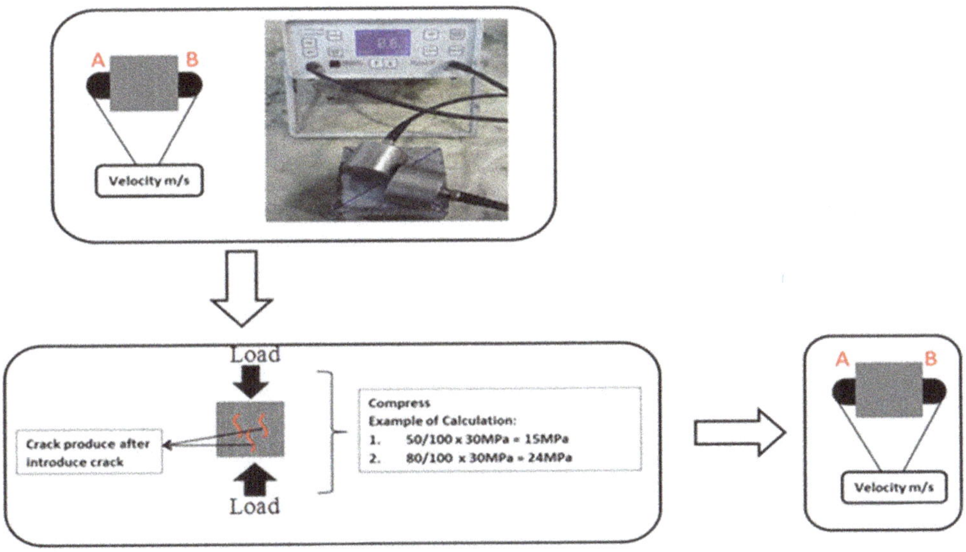

Table 1. Crack generated to epoxy-modified and normal mortar

Specimen	Cracks Generate	
Epoxy-modified Mortar	50% of maximum load	50EM
	80% of maximum load	80EM
Normal mortar	50% of maximum load	50NM
	80% of maximum load	80NM

as fifty epoxy-modified mortars (50EM), while the 80% one was denoted as the eighty epoxy-modified mortars (80EM). Similarly, the normal mortar specimens were denoted as fifty normal mortars (50NM) and eighty normal mortars (80NM). These are summarized in Table 1.

Compressive Strength

The specimen's ability to regain the initial compressive strength can be an indication of the self-healing process. The performance of epoxy resin without hardener as a self-healing agent has a potential to produce a maintenance-free mortar or self-healing mortar. The results are shown in Figures 2 to 4 for the

Figure 2. Compressive strength of self-healing epoxy modified and normal mortars after loaded at 28 days

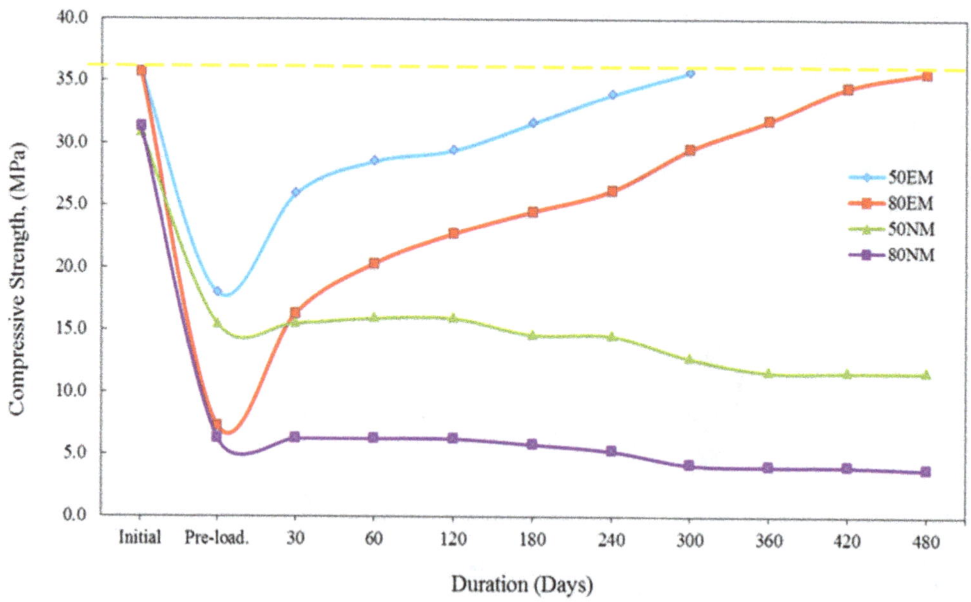

Figure 3. Compressive strength of self-healing epoxy modifies and normal mortars after loaded at 180 days

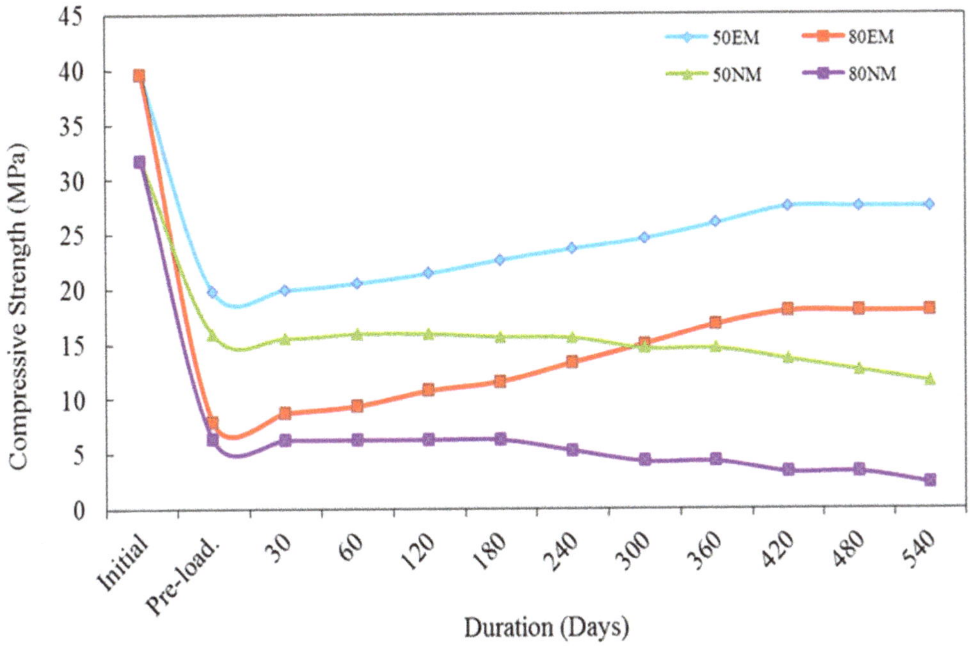

Figure 4. Compressive strength of self-healing epoxy modifies and normal mortars after loaded at 365 days

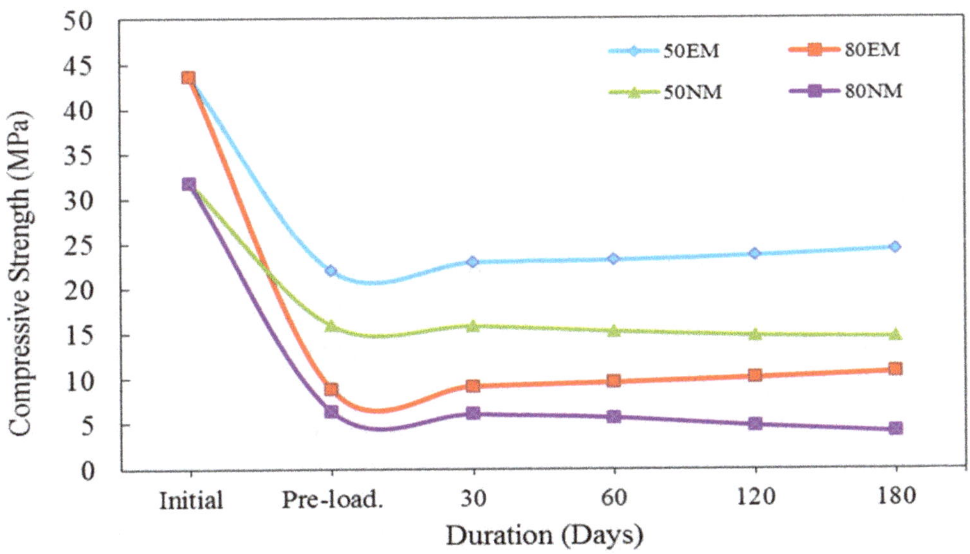

compressive strength of mortars after being loaded for 28 days, 180 days and 365 days. These figures showed the signs of compressive strength recovery in all cases. In the first month after the cracks were produced, the healing process was very fast, especially for epoxy-modified mortar cracked at 28 days. The initial compressive strength of epoxy-modified mortar was 36 MPa but after the crack was produced, the compressive strength decreased to 18 MPa for the 50EM while 7 MPa for the 80EM specimens. After 300 days of curing, the compressive strength of 50EM regained its initial strength while for the 80EM specimen, the compressive strength was 30 MPa. At 480 days of curing, the 80EM specimen was fully recovered and back to its original strength. That meant the crack inside the epoxy-modified mortar had been fully healed.

The compressive strength of epoxy-modified mortar which cracked at the age of 180 days showed a similar pattern but became constant towards the end of the testing period as shown in Figure 3. As shown in the Figure, the initial compressive strength after cracking was 19.8 MPa for the 50EM specimen, but increased to 20 MPa after a month of curing. On a similar note, 80EM specimens recorded 7.9 MPa of compressive strength after the production of crack and later it increased to 9 MPa after a month. However, the compressive strength of both specimens remained constant after 420 days of curing.

Similarly, the same pattern of results was observed in specimens cracked at the age of 365 days as shown in Figure 4. The initial compressive strength of epoxy-modified mortar was 45 MPa which reduced to 23 MPa for 50EM and less than 10 MPa for 80EM after the production of cracks. However, after 180 days of curing, the compressive strength for 50EM and 80EM increased to 24 MPa and 10.6 MPa, respectively. The increment of compressive strength was very small due to the age of the specimen. Some increment had shown which indicated that the good progress of self-healing function was occurring.

Generally, all the results showed an indication that the epoxy resin has reacted well with hydroxyl ions in repairing and healing the cracks inside the specimens without external intervention. The different healing rate was governed by the age of mortar. As explained, the young mortar heals faster than older mortar. The availability amount of unhardened epoxy resin to react with hydroxyl ions is also one of the factors that affect the healing rate. The epoxy resin shows that it can be used as a self-healing agent to heal the cracks automatically.

On the contrary, the compressive strength of normal mortar did not change and remained almost constant after a crack was produced. Even though the compressive strength was constant, it started to decrease after a certain time. This is probably due to the penetration of water and aggressive ions such as chloride ions, which deteriorate the mortars. As a result, the compressive strength was reduced and was found to deteriorate (Ariffin et al., 2015a).

The pattern of self-healing in terms of compressive strength for epoxy resin and bacteria is slightly different. The bacteria self-healing agent can repair the cracks faster than epoxy resin as reported by several other researchers, for example Bravo et al., (2015). In his research, the healing process can be as fast as three days with the immersion of specimen in the solution containing bacteria. The objective of self-healing without requiring any extra material and workers had been achieved by using epoxy resin without hardener as a healing agent in the mortar specimens.

Damage Degree of Concrete After Self-Healing

The degree of healing rate was calculated to evaluate the healing progress of epoxy-modified mortar. Figure 5 presents a trend of the degree of damage for 50EM and 80EM specimens after a certain period. The results showed a good self-healing performance especially in 50EM cracked at the age of 28

Figure 5. The damage of degree of self-healing epoxy modified mortar

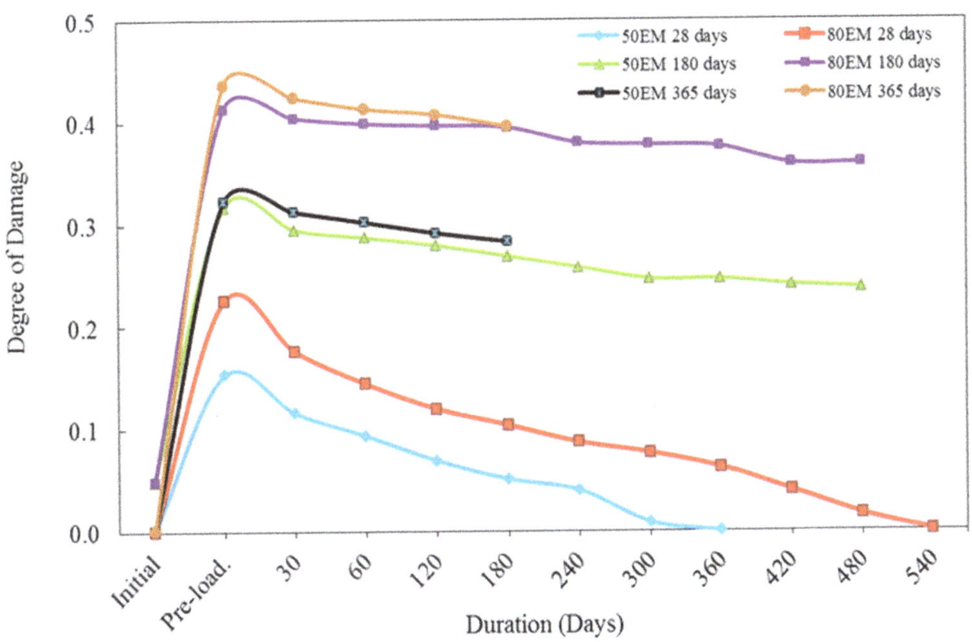

days, where the cracks were fully healed after 300 days of curing. The 80EM specimens produced crack at the age of 28 days took 480 days to completely heal the cracks. For epoxy-modified mortar cracked at the age of 180 days, the healing was slower and became constant after 300 days of curing. Only crack in 50EM was healed since 80EM had a limited amount of unhardened epoxy resin. For the specimens cracked at the age of 365 days, the healing was even slower. The point to emphasize is that the unhardened epoxy resin will still react with hydroxyl ions if no cracks are present to fill the pores. However, the progress of self-healing can be observed in the early ages after the crack was produced.

It was observed that earlier the damage occurred in the specimens, the better are the mechanical properties of the recovered material for the same degree of damage. The cement component and the epoxy resin matrix of the healing agent has led to a good adhesion, helping to heal the crack and reform its original state (Zhang and Rong, 2011). The results of self-healing performances, especially at 180 and 365 days, of producing crack showed a positive sign of healing the cracks automatically with the use of epoxy resin as a self-healing agent.

Permeability Test

The permeability test of the self-healing epoxy-modified mortar was carried out using the time for water to pass through the mortar. The water permeability time can be one of the indicators to evaluate the self-healing performance of epoxy resin in repairing the cracks. The time measurement of water permeability was recorded and presented in Figure 6. The time for water to pass through the self-healing epoxy-modified mortar indicates the number of pores and cracks inside the mortar.

It can be seen from the Figure that time taken for water to penetrate till the bottom of the specimen was initially 180 minutes. However, after the crack was produced, the time recorded was faster, which was 70 minutes to pass through for 50EM specimen and 30 minutes for 80EM specimen. The difference in the time is due to the different size of cracks, where 50EM specimen had smaller cracks and vice versa. As the mortar began to heal, the time taken for water to pass increased, which indicated that self-healing occurred inside the epoxy-modified mortar.

Figure 6. Time measurement for water permeability of mortar crack produce at 28 days

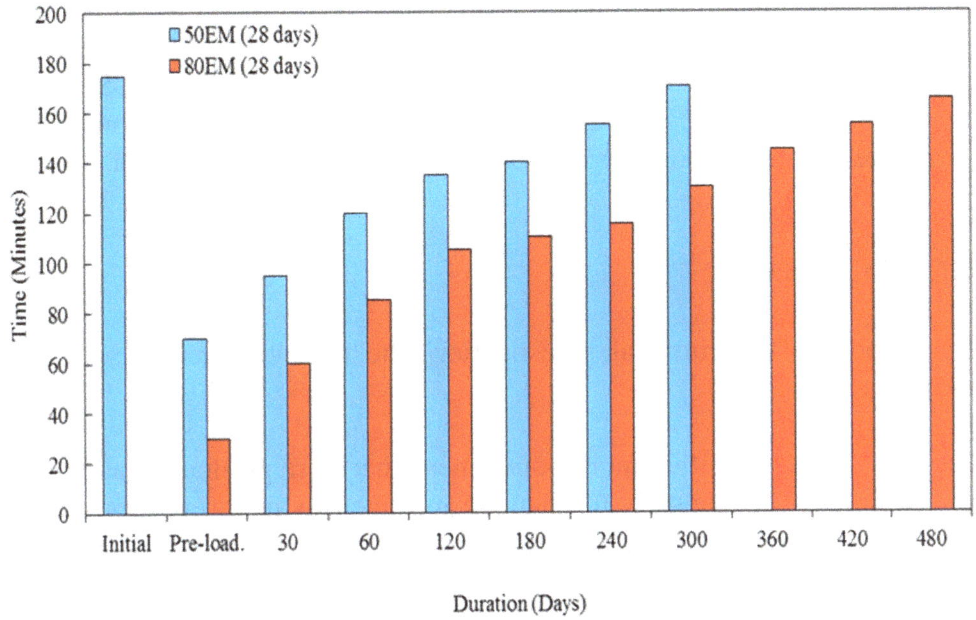

Reaction Between Epoxy Resin and Hydroxyl Ion

The motivation to carry out the study of using epoxy resin without hardener as a self-healing agent originated from the reaction between epoxy resin and hydroxyl ion. It is believed that the reaction rate can improve the properties of epoxy-modified mortar and act as a self-healing agent. The hydroxyl ion produced during cement hydration process is important for self-healing together with calcium-silicate-hydrate (CSH) gel. The hydroxyl ions are needed for self-healing mechanism to occur (Ariffin et al., 2015b). For the bacteria self-healing concrete, the bacteria will react with oxygen to produce calcite to precipitate and heal the cracks. The same mechanism occurred in epoxy-modified mortar where, when unhardened epoxy resin reacts with hydroxyl ions, it will harden and heal the cracks. Due to the elimination of hardener in the mortar, not all epoxy resin harden inside the mortar at an early age. The unhardened epoxy resin will react as a self-healing agent while the hardened epoxy resin contributes to the strength of the mortar specimens.

The self-healing mechanism is shown in the Scanning Electron Microscopy (SEM) morphology in Figure 7. This Figure showed that the cracks have been filled with the healing agent and the healing process started immediately after the cracks were developed. It also showed that the cracks had been filled with the reaction of epoxy resin and hydroxyl ion products. Besides, the CSH gel is also observed in Figure 7. The CSH gel is known as the main component that gives strength to the mortar produced from the hydration process of cement.

CONCLUSION

Based on the experimental investigation, the following remarks can be highlighted:

1. The epoxy-modified mortar specimens that were pre-loaded at 50% of ultimate load healed faster than the specimens which were subjected to 80% ultimate load.
2. The self-healing of epoxy-modified mortar, cracked at 28 days, completely healed after 300 days of curing.
3. The healing progress of cracks at age of 180 days of mortar showed a good performance where almost 60% of the cracks had been repaired.

Figure 7. The SEM morphology of self-healing reaction in epoxy-modified mortar

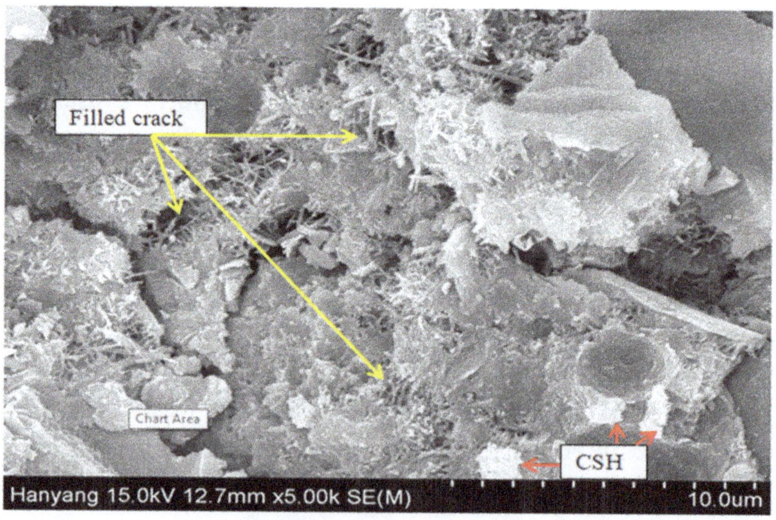

4. Water permeability of the self-healing epoxy-modified mortar showed an increase in the time taken by water penetration, which indicated that the cracks and pores inside the specimens had been repaired and filled by the reaction between epoxy resin and hydroxyl ions.

ACKNOWLEDGMENT

We, the editors greatly appreciate the exceptional contribution of Dr. Nur Farhayu Ariffin for carrying out this research and writing of the chapter. Indeed, members of her team, Dr. Khairunisa Muthusamy, Dr. Mohd Azreen Mohd Arifin, Dr. Nur Hafizah A. Khalid, Dr. Ghasan Fahim Huseien, and Dr. Mostafa Samadi deserve to be equally acknowledged for their co-authorship and immense collaboration in conducting/writing of the research.

REFERENCES

Ariffin, N. F., Hussin, M. W., Sam, A. R. M., Bhutta, M. A. R., Khalid, N. H. A., & Mirza, J. (2015b). Strength properties and molecular composition of epoxy-modified mortars. *Construction & Building Materials*, *94*, 315–322. doi:10.1016/j.conbuildmat.2015.06.056

Ariffin, N. F., Hussin, M. W., Sam, A. R. M., Lee, H. S., Khalid, N. H. A., Lim, N. H. A. S., & Samadi, M. (2015a). Mechanical properties and self-healing mechanism of epoxy mortar. *Jurnal Teknologi*, *77*(12), 37–44. doi:10.11113/jt.v77.6306

Bravo, F., Belie, N., De Boon, N., & Verstraete, W. (2015). Production of non-axenic ureolytic spores for self-healing concrete applications. *Construction & Building Materials*, *93*, 1034–1041. doi:10.1016/j.conbuildmat.2015.05.049

Edvardsen, C. (1999). Water permeability and autogenous healing of cracks in concrete. *ACI Materials Journal*, *96*(4), 448–454.

Gerilla, G. P., Teknomo, K., & Hokao, K. (2007). An environmental assessment of wood and steel reinforced concrete housing construction. *Building and Environment*, *42*(7), 2778–2784. doi:10.1016/j.buildenv.2006.07.021

Hager, M. D., Greil, P., Leyens, C., Van Der Zwaag, S., & Schubert, U. S. (2010). Self-healing materials. *Advanced Materials*, *22*(47), 5424–5430. doi:10.1002/adma.201003036 PMID:20839257

Jonkers, H. M., & Schlangen, E. (2008). Development of a bacteria-based self healing concrete. *Tailor Made Concrete Structures,* 425–430.

Jonkers, H. M., Thijssen, A., Muyzer, G., Copuroglu, O., & Schlangen, E. (2010). Application of bacteria as self-healing agent for the development of sustainable concrete. *Ecological Engineering*, *36*(2), 230–235. doi:10.1016/j.ecoleng.2008.12.036

Palin, D., Wiktor, V., & Jonkers, H. M. (2015). Autogenous healing of marine exposed concrete : Characterization and quanti fi cation through visual crack closure. *Cement and Concrete Research*, *73*, 17–24. doi:10.1016/j.cemconres.2015.02.021

Reinhardt, H. W., & Jooss, M. (2003). Permeability and self-healing of cracked concrete as a function of temperature and crack width. *Cement and Concrete Research*, *33*(7), 981–985. doi:10.1016/S0008-8846(02)01099-2

Stefan Jacobsen, S., & Sellevold, E. J. (1996). Self Healing of high strength concrete after deterioration by freeze / thaw. *Cement and Concrete Research*, *26*(I), 55–62. doi:10.1016/0008-8846(95)00179-4

Ter Heide, N. (2005). *Crack healing in hydrating concrete. Delft University of Technology*. Delft: Delft University of Technology.

Worrell, E., Price, L., Martin, N., Hendriks, C., & Meida, L. O. (2001). Carbon dioxide emissions from the global cement industry. *Annual Review of Energy and the Environment*, *26*(1), 303–329. doi:10.1146/annurev.energy.26.1.303

Zhang, M., Xing, F., Shi, K. Y., & Liu, P. (2011). Novel self-healing techniques for cement matrix microcrack. *Advanced Materials Research*, *239-242*, 3310–3313. doi:10.4028/www.scientific.net/AMR.239-242.3310

Zhang, M. Q., & Rong, M. Z. (2011). Extrinsic self - healing via addition polymerization. Self-Healing Polymers and Polymer Composites, 111–166.

Zhong, W., & Yao, W. (2008). Influence of damage degree on self-healing of concrete. *Construction & Building Materials*, *22*(6), 1137–1142. doi:10.1016/j.conbuildmat.2007.02.006

Related Readings

To continue IGI Global's long-standing tradition of advancing innovation through emerging research, please find below a compiled list of recommended IGI Global book chapters and journal articles in the areas of concrete construction, recycled waste materials, and construction materials. These related readings will provide additional information and guidance to further enrich your knowledge and assist you with your own research.

Afonso, J. M. (2018). The Power of Monsanto's Stone: Contribution to the Study of the Sustainable Adaptative Strategies. In I. Rosa, J. Lopes, R. Ribeiro, & A. Mendes (Eds.), *Handbook of Research on Methods and Tools for Assessing Cultural Landscape Adaptation* (pp. 122–152). Hershey, PA: IGI Global. doi:10.4018/978-1-5225-4186-8.ch006

Akpinar, E., & Ersin, S. (2017). A Comparative Investigation of Structural Performance of Typical and Non-Ducitle Public RC Buildings Strengthened Using Friction Dampers and RC Walls. In V. Plevris, G. Kremmyda, & Y. Fahjan (Eds.), *Performance-Based Seismic Design of Concrete Structures and Infrastructures* (pp. 112–127). Hershey, PA: IGI Global. doi:10.4018/978-1-5225-2089-4.ch005

Ali, S., & Murari, V. (2019). Miniaturization of Test Specimen for Composites. In M. Uthayakumar, S. Raj, T. Ko, S. Kumaran, & J. Davim (Eds.), *Handbook of Research on Green Engineering Techniques for Modern Manufacturing* (pp. 49–76). Hershey, PA: IGI Global. doi:10.4018/978-1-5225-5445-5.ch004

Amador, H. C. (2018). Contemporary Urbanizations in Public Water Reservoirs: Floating Villages of Alqueva Reservoir. In I. Rosa, J. Lopes, R. Ribeiro, & A. Mendes (Eds.), *Handbook of Research on Methods and Tools for Assessing Cultural Landscape Adaptation* (pp. 249–275). Hershey, PA: IGI Global. doi:10.4018/978-1-5225-4186-8.ch010

Arumugaprabu, V. R., D. J., & R., P. (2019). Effective Utilization of Industrial Wastes for Preparing Polymer Matrix Composites: Usage of Industrial Wastes. In M. Uthayakumar, S. Raj, T. Ko, S. Kumaran, & J. Davim (Eds.), Handbook of Research on Green Engineering Techniques for Modern Manufacturing (pp. 250-261). Hershey, PA: IGI Global. doi:10.4018/978-1-5225-5445-5.ch013

Atik, M., Sadek, M., & Shahrour, I. (2017). Single-Run Adaptive Pushover Procedure for Shear Wall Structures. In V. Plevris, G. Kremmyda, & Y. Fahjan (Eds.), *Performance-Based Seismic Design of Concrete Structures and Infrastructures* (pp. 59–83). Hershey, PA: IGI Global. doi:10.4018/978-1-5225-2089-4.ch003

Balsas, C. J. (2018). A Novel Approach to Studying Cultural Landscapes at the Watershed Level. In I. Rosa, J. Lopes, R. Ribeiro, & A. Mendes (Eds.), *Handbook of Research on Methods and Tools for Assessing Cultural Landscape Adaptation* (pp. 221–248). Hershey, PA: IGI Global. doi:10.4018/978-1-5225-4186-8.ch009

Baptista, J. L., Ribeiro, J. T., & Henriques, C. D. (2018). Construction Information Map: Support for Sustainable Architecture Projects in Developing Countries – Angola Case Study. In I. Rosa, J. Lopes, R. Ribeiro, & A. Mendes (Eds.), *Handbook of Research on Methods and Tools for Assessing Cultural Landscape Adaptation* (pp. 323–345). Hershey, PA: IGI Global. doi:10.4018/978-1-5225-4186-8.ch013

Bathrinath, S., Saranyadevi, S., Thirumalai Kumaran, S., & Saravanasankar, S. (2019). PageRank Algorithm-Based Recommender System Using Uniformly Average Rating Matrix. In M. Uthayakumar, S. Raj, T. Ko, S. Kumaran, & J. Davim (Eds.), *Handbook of Research on Green Engineering Techniques for Modern Manufacturing* (pp. 99–112). Hershey, PA: IGI Global. doi:10.4018/978-1-5225-5445-5.ch006

Bobo, T., & Nechena, H. (2018). Fighting for Existence and Recognition Among Sub-Dynasty Communities: A Case Study of the Nerumedzo People in Zimbabwe. In I. Rosa, J. Lopes, R. Ribeiro, & A. Mendes (Eds.), *Handbook of Research on Methods and Tools for Assessing Cultural Landscape Adaptation* (pp. 53–71). Hershey, PA: IGI Global. doi:10.4018/978-1-5225-4186-8.ch003

Boonkerd, K. (2017). Development and Modification of Natural Rubber for Advanced Application. In T. Kobayashi (Ed.), *Applied Environmental Materials Science for Sustainability* (pp. 44–76). Hershey, PA: IGI Global. doi:10.4018/978-1-5225-1971-3.ch003

Boonmahitthisud, A. (2017). Natural Rubber and Rubber Blend Nanocomposites: Reinforcement of Natural Rubber with Polymer-Encapsulated Inorganic Nanohybrid Particles. In T. Kobayashi (Ed.), *Applied Environmental Materials Science for Sustainability* (pp. 77–105). Hershey, PA: IGI Global. doi:10.4018/978-1-5225-1971-3.ch004

Borowski, J. (2018). Impact of Climbing Plants on Buildings and Their Environment. In E. Rynska, U. Kozminska, K. Zinowiec-Cieplik, J. Rucinska, & B. Szybinska-Matusiak (Eds.), *Design Solutions for nZEB Retrofit Buildings* (pp. 297–309). Hershey, PA: IGI Global. doi:10.4018/978-1-5225-4105-9.ch013

Brownell, B. E. (2019). Determining Architecture's Footprint: Preliminary Methods for Measuring the True Environmental Impact of Buildings. In G. Koç & B. Christiansen (Eds.), *Reusable and Sustainable Building Materials in Modern Architecture* (pp. 28–59). Hershey, PA: IGI Global. doi:10.4018/978-1-5225-6995-4.ch002

Carrillo, K. L., & Kobayashi, T. (2017). Natural Material Source of Bagasse Cellulose and Their Application to Hydrogel Films. In T. Kobayashi (Ed.), *Applied Environmental Materials Science for Sustainability* (pp. 19–43). Hershey, PA: IGI Global. doi:10.4018/978-1-5225-1971-3.ch002

Clementi, F., Di Sciascio, G., Di Sciascio, S., & Lenci, S. (2017). Influence of the Shear-Bending Interaction on the Global Capacity of Reinforced Concrete Frames: A Brief Overview of the New Perspectives. In V. Plevris, G. Kremmyda, & Y. Fahjan (Eds.), *Performance-Based Seismic Design of Concrete Structures and Infrastructures* (pp. 84–111). Hershey, PA: IGI Global. doi:10.4018/978-1-5225-2089-4.ch004

Cui, X., Zeng, S., Li, Z., Zheng, Q., Yu, X., & Han, B. (2018). Advanced Composites for Civil Engineering Infrastructures. In K. Kumar & J. Davim (Eds.), *Composites and Advanced Materials for Industrial Applications* (pp. 212–248). Hershey, PA: IGI Global. doi:10.4018/978-1-5225-5216-1.ch010

Dixit, A. (2018). Application of Silica-Gel-Reinforced Aluminium Composite on the Piston of Internal Combustion Engine: Comparative Study of Silica-Gel-Reinforced Aluminium Composite Piston With Aluminium Alloy Piston. In K. Kumar & J. Davim (Eds.), *Composites and Advanced Materials for Industrial Applications* (pp. 63–98). Hershey, PA: IGI Global. doi:10.4018/978-1-5225-5216-1.ch004

Elsayed, A. M., Dakkama, H. J., Mahmoud, S., Al-Dadah, R., & Kaialy, W. (2017). Sustainable Cooling Research Using Activated Carbon Adsorbents and Their Environmental Impact. In T. Kobayashi (Ed.), *Applied Environmental Materials Science for Sustainability* (pp. 186–221). Hershey, PA: IGI Global. doi:10.4018/978-1-5225-1971-3.ch009

Fahjan, Y. M., Kara, F. İ., & Mert, A. (2017). Selection and Scaling Time History Records for Performance-Based Design. In V. Plevris, G. Kremmyda, & Y. Fahjan (Eds.), *Performance-Based Seismic Design of Concrete Structures and Infrastructures* (pp. 1–35). Hershey, PA: IGI Global. doi:10.4018/978-1-5225-2089-4.ch001

Faisal, N., & Kumar, K. (2019). Recycling and Reuse of Building Materials From Construction and Demolition: An Environmental Evaluation for Sustainable Growth. In G. Koç & B. Christiansen (Eds.), *Reusable and Sustainable Building Materials in Modern Architecture* (pp. 60–79). Hershey, PA: IGI Global. doi:10.4018/978-1-5225-6995-4.ch003

Fernando, P. R., Hamigah, T., Disne, S., Wickramasingha, G. G., & Sutharshan, A. (2018). The Evaluation of Engineering Properties of Low Cost Concrete Blocks by Partial Doping of Sand with Sawdust: Low Cost Sawdust Concrete Block. *International Journal of Strategic Engineering, 1*(2), 26–42. doi:10.4018/IJoSE.2018070103

Firląg, S. (2018). Definition of nZEB Renovation Standard. In E. Rynska, U. Kozminska, K. Zinowiec-Cieplik, J. Rucinska, & B. Szybinska-Matusiak (Eds.), *Design Solutions for nZEB Retrofit Buildings* (pp. 1–23). Hershey, PA: IGI Global. doi:10.4018/978-1-5225-4105-9.ch001

Golara, A. (2017). A Framework and Case Study for the Resilience of Infrastructures. In V. Plevris, G. Kremmyda, & Y. Fahjan (Eds.), *Performance-Based Seismic Design of Concrete Structures and Infrastructures* (pp. 261–274). Hershey, PA: IGI Global. doi:10.4018/978-1-5225-2089-4.ch010

Gonçalves, C. (2018). Models, Methods, and Metrics to Measure Socioeconomic Resilience: Two Portuguese Urban Systems as Case Studies. In I. Rosa, J. Lopes, R. Ribeiro, & A. Mendes (Eds.), *Handbook of Research on Methods and Tools for Assessing Cultural Landscape Adaptation* (pp. 346–367). Hershey, PA: IGI Global. doi:10.4018/978-1-5225-4186-8.ch014

Gorgulu, O., & Taskin, B. (2017). A Review of the Accuracy of Force- and Deformation-Based Methods in Determining the Seismic Capacity of Rehabilitated RC School Buildings. In V. Plevris, G. Kremmyda, & Y. Fahjan (Eds.), *Performance-Based Seismic Design of Concrete Structures and Infrastructures* (pp. 172–196). Hershey, PA: IGI Global. doi:10.4018/978-1-5225-2089-4.ch007

Guzmán, A. M., Maldonado, N. G., & Affranchino, G. (2019). Comprehensive Evaluation for Mortars and Concretes Incorporating Wastes. In G. Koç & B. Christiansen (Eds.), *Reusable and Sustainable Building Materials in Modern Architecture* (pp. 108–136). Hershey, PA: IGI Global. doi:10.4018/978-1-5225-6995-4.ch006

Horta, M. B., Cabral, M. I., Pires, I., Bachi, L. S., Luz, A., Fernandes, G. W., ... Carvalho-Ribeiro, S. (2018). Assessing Urban Ecosystem Services: Different Methodological Approaches Applied in Brazil, Germany, and Portugal. In I. Rosa, J. Lopes, R. Ribeiro, & A. Mendes (Eds.), *Handbook of Research on Methods and Tools for Assessing Cultural Landscape Adaptation* (pp. 183–220). Hershey, PA: IGI Global. doi:10.4018/978-1-5225-4186-8.ch008

Hukkalainen, M. A., Klobut, K., Mäkeläinen, T., Dimitriou, V., & Leszczyński, D. (2018). Improving the Design of Energy-Efficient Building Retrofitting: Design Guidelines, Energy Simulations, and Selecting of Technologies. In E. Rynska, U. Kozminska, K. Zinowiec-Cieplik, J. Rucinska, & B. Szybinska-Matusiak (Eds.), *Design Solutions for nZEB Retrofit Buildings* (pp. 165–185). Hershey, PA: IGI Global. doi:10.4018/978-1-5225-4105-9.ch007

Huyen, P. T. (2017). Clay Minerals Converted to Porous Materials and Their Application: Challenge and Perspective. In T. Kobayashi (Ed.), *Applied Environmental Materials Science for Sustainability* (pp. 141–164). Hershey, PA: IGI Global. doi:10.4018/978-1-5225-1971-3.ch007

Ida, A. (2019). Energetic Forms of Matter. In G. Koç & B. Christiansen (Eds.), *Reusable and Sustainable Building Materials in Modern Architecture* (pp. 137–165). Hershey, PA: IGI Global. doi:10.4018/978-1-5225-6995-4.ch007

Ikematsu, S., Tada, I., & Nagasaki, Y. (2017). Analysis of Lipids Produced by Microalgae Isolated from the Area around Okinawa, Japan. In T. Kobayashi (Ed.), *Applied Environmental Materials Science for Sustainability* (pp. 222–233). Hershey, PA: IGI Global. doi:10.4018/978-1-5225-1971-3.ch010

Ion, I. Ă., & Costel, S. (2016). Management Efficiency Building Materials. *International Journal of Sustainable Economies Management*, *5*(2), 14–20. doi:10.4018/IJSEM.2016040102

Jayapalan, S. (2018). A Review of Chemical Treatments on Natural Fibers-Based Hybrid Composites for Engineering Applications. In K. Kumar & J. Davim (Eds.), *Composites and Advanced Materials for Industrial Applications* (pp. 16–37). Hershey, PA: IGI Global. doi:10.4018/978-1-5225-5216-1.ch002

Jędrzejuk, H. I. (2018). Selection of Design Methods in the Modernization Process of Buildings to the nZEB Standard. In E. Rynska, U. Kozminska, K. Zinowiec-Cieplik, J. Rucinska, & B. Szybinska-Matusiak (Eds.), *Design Solutions for nZEB Retrofit Buildings* (pp. 24–68). Hershey, PA: IGI Global. doi:10.4018/978-1-5225-4105-9.ch002

Jędrzejuk, H. I. (2018). Selection of Renewable Energy Sources for Buildings. In E. Rynska, U. Kozminska, K. Zinowiec-Cieplik, J. Rucinska, & B. Szybinska-Matusiak (Eds.), *Design Solutions for nZEB Retrofit Buildings* (pp. 69–97). Hershey, PA: IGI Global. doi:10.4018/978-1-5225-4105-9.ch003

Kandasamy, J., & Sakthivel, A. R. K. E. K., V., M. S., S., M. T. H., S., Reddy, V. S., & Gutha, J. (2019). Assessment of Remanufacturability Index for an Automotive Product: A Case Study. In M. Uthayakumar, S. Raj, T. Ko, S. Kumaran, & J. Davim (Eds.), Handbook of Research on Green Engineering Techniques for Modern Manufacturing (pp. 296-308). Hershey, PA: IGI Global. doi:10.4018/978-1-5225-5445-5.ch016

Kandasamy, J., & Sakthivel, A. R. K. E. K., V., Reddy, V. S., & K. S., B. (2019). Application of Cluster Analysis for Identifying Potential Automotive Organizations Towards the Conduct of Green Manufacturing Sustainability Studies. In M. Uthayakumar, S. Raj, T. Ko, S. Kumaran, & J. Davim (Eds.), Handbook of Research on Green Engineering Techniques for Modern Manufacturing (pp. 309-322). Hershey, PA: IGI Global. doi:10.4018/978-1-5225-5445-5.ch017

Katrancioglu, S., Kilic, H. S., & Uslu, C. A. (2019). Solution Approaches for Reverse Logistics Considering Recovery Options: A Literature Review. In M. Uthayakumar, S. Raj, T. Ko, S. Kumaran, & J. Davim (Eds.), *Handbook of Research on Green Engineering Techniques for Modern Manufacturing* (pp. 192–210). Hershey, PA: IGI Global. doi:10.4018/978-1-5225-5445-5.ch011

Klimowicz, J. (2018). Chosen Case Studies of nZEB Retrofit Buildings. In E. Rynska, U. Kozminska, K. Zinowiec-Cieplik, J. Rucinska, & B. Szybinska-Matusiak (Eds.), *Design Solutions for nZEB Retrofit Buildings* (pp. 209–227). Hershey, PA: IGI Global. doi:10.4018/978-1-5225-4105-9.ch009

Kobayashi, T. (2017). Introduction of Environmental Materials. In T. Kobayashi (Ed.), *Applied Environmental Materials Science for Sustainability* (pp. 1–18). Hershey, PA: IGI Global. doi:10.4018/978-1-5225-1971-3.ch001

Kozminska, U., & Rynska, E. D. (2018). Existing Buildings: How to Meet an nZeb Standard – The Architect's Perspective. In E. Rynska, U. Kozminska, K. Zinowiec-Cieplik, J. Rucinska, & B. Szybinska-Matusiak (Eds.), *Design Solutions for nZEB Retrofit Buildings* (pp. 228–247). Hershey, PA: IGI Global. doi:10.4018/978-1-5225-4105-9.ch010

Kumar, M., Kumar, R., Kumar, S., & Prakash, C. (2019). Biomechanical Properties of Orthopedic and Dental Implants: A Comprehensive Review. In M. Uthayakumar, S. Raj, T. Ko, S. Kumaran, & J. Davim (Eds.), *Handbook of Research on Green Engineering Techniques for Modern Manufacturing* (pp. 1–13). Hershey, PA: IGI Global. doi:10.4018/978-1-5225-5445-5.ch001

Kumari, N., & Kumar, K. (2018). Fabrication of Orthotic Calipers With Epoxy-Based Green Composite. In K. Kumar & J. Davim (Eds.), *Composites and Advanced Materials for Industrial Applications* (pp. 157–176). Hershey, PA: IGI Global. doi:10.4018/978-1-5225-5216-1.ch008

Kuppusamy, R. R. (2018). Development of Aerospace Composite Structures Through Vacuum-Enhanced Resin Transfer Moulding Technology (VERTMTy): Vacuum-Enhanced Resin Transfer Moulding. In K. Kumar & J. Davim (Eds.), *Composites and Advanced Materials for Industrial Applications* (pp. 99–111). Hershey, PA: IGI Global. doi:10.4018/978-1-5225-5216-1.ch005

Lavrador, A., & Rocha, J. (2018). The Role of Landscape in the Representation of Portuguese Wine Producing Regions. In I. Rosa, J. Lopes, R. Ribeiro, & A. Mendes (Eds.), *Handbook of Research on Methods and Tools for Assessing Cultural Landscape Adaptation* (pp. 276–298). Hershey, PA: IGI Global. doi:10.4018/978-1-5225-4186-8.ch011

Li, K., & Kobayashi, T. (2017). Ionic Liquids and Poly (Ionic Liquid)s Used as Green Solvent and Ultrasound Responded Materials. In T. Kobayashi (Ed.), *Applied Environmental Materials Science for Sustainability* (pp. 327–346). Hershey, PA: IGI Global. doi:10.4018/978-1-5225-1971-3.ch015

Lourenço, A. C. (2018). City Landscape: Confluence Between Ecological Conditions and Urban Morphology in the City of Lisbon. In I. Rosa, J. Lopes, R. Ribeiro, & A. Mendes (Eds.), *Handbook of Research on Methods and Tools for Assessing Cultural Landscape Adaptation* (pp. 368–395). Hershey, PA: IGI Global. doi:10.4018/978-1-5225-4186-8.ch015

Lourenço, E. J., Moita, N., Esteves, S., Peças, P., Ribeiro, I. I., Pereira, J. P., & Oliveira, L. M. (2019). Multi-Perspective Eco-Efficiency Assessment to Foster Sustainability in Plastic Parts Production: An Integrated Tool for Industrial Use. In M. Uthayakumar, S. Raj, T. Ko, S. Kumaran, & J. Davim (Eds.), *Handbook of Research on Green Engineering Techniques for Modern Manufacturing* (pp. 212–249). Hershey, PA: IGI Global. doi:10.4018/978-1-5225-5445-5.ch012

Louro, M., & Oliveira, F. (2018). In Search of a Lost Authenticity: Tourism, Renaturation, and City – The Case of the Tagus Estuary. In I. Rosa, J. Lopes, R. Ribeiro, & A. Mendes (Eds.), *Handbook of Research on Methods and Tools for Assessing Cultural Landscape Adaptation* (pp. 72–89). Hershey, PA: IGI Global. doi:10.4018/978-1-5225-4186-8.ch004

Loy, J., & Schork, T. (2019). Building Relationships: Changing Technology and Society. In G. Koç & B. Christiansen (Eds.), *Reusable and Sustainable Building Materials in Modern Architecture* (pp. 166–187). Hershey, PA: IGI Global. doi:10.4018/978-1-5225-6995-4.ch008

M., S. M., & V., A. (2019). Kinematic Modelling and Simulation of 8 Degrees of Freedom SCARA Robot. In M. Uthayakumar, S. Raj, T. Ko, S. Kumaran, & J. Davim (Eds.), *Handbook of Research on Green Engineering Techniques for Modern Manufacturing* (pp. 77-97). Hershey, PA: IGI Global. doi:10.4018/978-1-5225-5445-5.ch005

Mahmood, W. A., & Azarian, M. H. (2017). Inorganic-Organic Composite Materials from Liquid Natural Rubber and Epoxidised Natural Rubber Derivatives: Prospects and Applications. In T. Kobayashi (Ed.), *Applied Environmental Materials Science for Sustainability* (pp. 128–140). Hershey, PA: IGI Global. doi:10.4018/978-1-5225-1971-3.ch006

Marzouk, M., & Abdelkader, E. M. (2017). Minimizing Construction Emissions Using Building Information Modeling and Decision-Making Techniques. *International Journal of 3-D Information Modeling, 6*(2), 14-35. doi:10.4018/IJ3DIM.2017040102

Marzouk, M., & El-Razek, M. A. (2017). Selecting Demolition Waste Materials Disposal Alternatives Using Fuzzy TOPSIS Technique. *International Journal of Natural Computing Research, 6*(2), 38–57. doi:10.4018/IJNCR.2017070103

McIntosh, J., Marques, B., & Hatton, W. (2018). Indigenous Cultural Knowledge for Therapeutic Landscape Design. In I. Rosa, J. Lopes, R. Ribeiro, & A. Mendes (Eds.), *Handbook of Research on Methods and Tools for Assessing Cultural Landscape Adaptation* (pp. 28–52). Hershey, PA: IGI Global. doi:10.4018/978-1-5225-4186-8.ch002

Melkumyan, M. G. (2017). Original and Innovative Structural Concepts for Design, Non-Linear Analysis, and Construction of Multi-Story Base Isolated Buildings. In V. Plevris, G. Kremmyda, & Y. Fahjan (Eds.), *Performance-Based Seismic Design of Concrete Structures and Infrastructures* (pp. 197–238). Hershey, PA: IGI Global. doi:10.4018/978-1-5225-2089-4.ch008

Mukhopadhyay, A., Barman, T. K., & Sahoo, P. (2018). Electroless Nickel Coatings for High Temperature Applications. In K. Kumar & J. Davim (Eds.), *Composites and Advanced Materials for Industrial Applications* (pp. 297–331). Hershey, PA: IGI Global. doi:10.4018/978-1-5225-5216-1.ch013

Munir, M. J., Kazmi, S. M., Wu, Y., & Patnaikuni, I. (2018). A Literature Review on Alkali Silica Reactivity of Concrete: Consequences and Challenges. *International Journal of Strategic Engineering, 1*(2), 43–62. doi:10.4018/IJoSE.2018070104

Náprstek, J., & Fischer, C. (2017). Dynamic Stability and Post-Critical Processes of Slender Auto-Parametric Systems. In V. Plevris, G. Kremmyda, & Y. Fahjan (Eds.), *Performance-Based Seismic Design of Concrete Structures and Infrastructures* (pp. 128–171). Hershey, PA: IGI Global. doi:10.4018/978-1-5225-2089-4.ch006

O'Donnell, C., & Pranger, D. (2019). Rethinking Waste Through Design. In G. Koç & B. Christiansen (Eds.), *Reusable and Sustainable Building Materials in Modern Architecture* (pp. 93–107). Hershey, PA: IGI Global. doi:10.4018/978-1-5225-6995-4.ch005

Oliveira, M. D., & Ribeiro, J. T. (2018). Industrial Heritage as an Operative Territorial Resource: Cultural Landscape of Alentejo Pyrite. In I. Rosa, J. Lopes, R. Ribeiro, & A. Mendes (Eds.), *Handbook of Research on Methods and Tools for Assessing Cultural Landscape Adaptation* (pp. 153–181). Hershey, PA: IGI Global. doi:10.4018/978-1-5225-4186-8.ch007

Oliveira, M. R., Jorge, D., & Peças, P. (2019). Methodology of Operationalization of KPIs for Shop-Floor. In M. Uthayakumar, S. Raj, T. Ko, S. Kumaran, & J. Davim (Eds.), *Handbook of Research on Green Engineering Techniques for Modern Manufacturing* (pp. 163–191). Hershey, PA: IGI Global. doi:10.4018/978-1-5225-5445-5.ch010

Onutai, S., Jiemsirilers, S., & Kobayashi, T. (2017). Geopolymer Sourced with Fly Ash and Industrial Aluminum Waste for Sustainable Materials. In T. Kobayashi (Ed.), *Applied Environmental Materials Science for Sustainability* (pp. 165–185). Hershey, PA: IGI Global. doi:10.4018/978-1-5225-1971-3.ch008

Pakdamar, F. (2017). Fuzzy Logic Applications for Performance-Based Design. In V. Plevris, G. Kremmyda, & Y. Fahjan (Eds.), *Performance-Based Seismic Design of Concrete Structures and Infrastructures* (pp. 239–260). Hershey, PA: IGI Global. doi:10.4018/978-1-5225-2089-4.ch009

Pierzchalski, M. (2018). Single-Family Residential Building Energy Retrofit: A Case Study. In E. Rynska, U. Kozminska, K. Zinowiec-Cieplik, J. Rucinska, & B. Szybinska-Matusiak (Eds.), *Design Solutions for nZEB Retrofit Buildings* (pp. 248–274). Hershey, PA: IGI Global. doi:10.4018/978-1-5225-4105-9.ch011

Potiyaraj, P. (2017). Poly (Lactic Acid) Generated for Advanced Materials. In T. Kobayashi (Ed.), *Applied Environmental Materials Science for Sustainability* (pp. 106–127). Hershey, PA: IGI Global. doi:10.4018/978-1-5225-1971-3.ch005

Prakash, C., Singh, S., Abdul-Rani, A. M., Uddin, M. S., Pabla, B. S., & Puri, S. (2019). Spark Plasma Sintering of Mg-Zn-Mn-Si-HA Alloy for Bone Fixation Devices: Fabrication of Biodegradable Low Elastic Porous Mg-Zn-Mn-Si-HA Alloy. In M. Uthayakumar, S. Raj, T. Ko, S. Kumaran, & J. Davim (Eds.), *Handbook of Research on Green Engineering Techniques for Modern Manufacturing* (pp. 282–295). Hershey, PA: IGI Global. doi:10.4018/978-1-5225-5445-5.ch015

Puglisi, V., & Ciaramella, A. (2019). Urban Quality Assessment at the Neighborhood Scale: An Experimental Approach. In G. Koç & B. Christiansen (Eds.), *Reusable and Sustainable Building Materials in Modern Architecture* (pp. 188–220). Hershey, PA: IGI Global. doi:10.4018/978-1-5225-6995-4.ch009

R., K., Tambrallimath, V., Kuppahalli, P., & N., S. (2019). Additive Manufacturing Process and Their Applications for Green Technology. In M. Uthayakumar, S. Raj, T. Ko, S. Kumaran, & J. Davim (Eds.), *Handbook of Research on Green Engineering Techniques for Modern Manufacturing* (pp. 262-281). Hershey, PA: IGI Global. doi:10.4018/978-1-5225-5445-5.ch014

Ramdani, N., & Azibi, M. (2018). Polymer Composite Materials for Microelectronics Packaging Applications: Composites for Microelectronics Packaging. In K. Kumar & J. Davim (Eds.), *Composites and Advanced Materials for Industrial Applications* (pp. 177–211). Hershey, PA: IGI Global. doi:10.4018/978-1-5225-5216-1.ch009

Ribeiro, R. J., Lopes, J. C., & Boucault, F. (2018). Alternative Tool for an Integrative Landscape Interpretation: Case Study of the Arrábida Maritime Coast, Portugal. In I. Rosa, J. Lopes, R. Ribeiro, & A. Mendes (Eds.), *Handbook of Research on Methods and Tools for Assessing Cultural Landscape Adaptation* (pp. 299–322). Hershey, PA: IGI Global. doi:10.4018/978-1-5225-4186-8.ch012

Rucińska, J. (2018). Energy Simulations as a Tool in Integrated Design Process. In E. Rynska, U. Kozminska, K. Zinowiec-Cieplik, J. Rucinska, & B. Szybinska-Matusiak (Eds.), *Design Solutions for nZEB Retrofit Buildings* (pp. 141–164). Hershey, PA: IGI Global. doi:10.4018/978-1-5225-4105-9.ch006

Rucińska, J. (2018). Improving the Energy Quality and Indoor Environmental Quality in Retrofit Buildings. In E. Rynska, U. Kozminska, K. Zinowiec-Cieplik, J. Rucinska, & B. Szybinska-Matusiak (Eds.), *Design Solutions for nZEB Retrofit Buildings* (pp. 186–208). Hershey, PA: IGI Global. doi:10.4018/978-1-5225-4105-9.ch008

Rynska, E. D. (2018). 3XE: Efficiency, Ecosphere, Economics. In E. Rynska, U. Kozminska, K. Zinowiec-Cieplik, J. Rucinska, & B. Szybinska-Matusiak (Eds.), *Design Solutions for nZEB Retrofit Buildings* (pp. 98–114). Hershey, PA: IGI Global. doi:10.4018/978-1-5225-4105-9.ch004

Sahoo, S. (2018). Laminated Composite Hypar Shells as Roofing Units: Static and Dynamic Behavior. In K. Kumar & J. Davim (Eds.), *Composites and Advanced Materials for Industrial Applications* (pp. 249–269). Hershey, PA: IGI Global. doi:10.4018/978-1-5225-5216-1.ch011

Sanal, I. (2017). A Review on Reduced Environmental Impacts of Alternative Green Concrete Productions. *International Journal of Public and Private Perspectives on Healthcare, Culture, and the Environment, 1*(2), 55–68. doi:10.4018/IJPPPHCE.2017070104

Satoh, M. (2017). Metal Ion Separation with Functional Adsorbents and Phytoremediation Used as Sustainable Technologies. In T. Kobayashi (Ed.), *Applied Environmental Materials Science for Sustainability* (pp. 284–312). Hershey, PA: IGI Global. doi:10.4018/978-1-5225-1971-3.ch013

Sezgin, H., & Berkalp, O. B. (2018). Textile-Reinforced Composites for the Automotive Industry. In K. Kumar & J. Davim (Eds.), *Composites and Advanced Materials for Industrial Applications* (pp. 129–156). Hershey, PA: IGI Global. doi:10.4018/978-1-5225-5216-1.ch007

Shah, M. Z., Gazder, U., Bhatti, M. S., & Hussain, M. (2018). Comparative Performance Evaluation of Effects of Modifier in Asphaltic Concrete Mix. *International Journal of Strategic Engineering, 1*(2), 13–25. doi:10.4018/IJoSE.2018070102

Sharma, N., & Kumar, K. (2018). Fabrication of Porous NiTi Alloy Using Organic Binders. In K. Kumar & J. Davim (Eds.), *Composites and Advanced Materials for Industrial Applications* (pp. 38–62). Hershey, PA: IGI Global. doi:10.4018/978-1-5225-5216-1.ch003

Silva, C. F. (2018). Time Operations. In I. Rosa, J. Lopes, R. Ribeiro, & A. Mendes (Eds.), *Handbook of Research on Methods and Tools for Assessing Cultural Landscape Adaptation* (pp. 90–121). Hershey, PA: IGI Global. doi:10.4018/978-1-5225-4186-8.ch005

Silva, C. F., & Fernandes, M. C. (2018). The Cycles of Impermanent Alterity in Nazaré. In I. Rosa, J. Lopes, R. Ribeiro, & A. Mendes (Eds.), *Handbook of Research on Methods and Tools for Assessing Cultural Landscape Adaptation* (pp. 434–465). Hershey, PA: IGI Global. doi:10.4018/978-1-5225-4186-8.ch018

Singh, R., & Dutta, S. (2018). Visible Light Active Nanocomposites for Photocatalytic Applications. In K. Kumar & J. Davim (Eds.), *Composites and Advanced Materials for Industrial Applications* (pp. 270–296). Hershey, PA: IGI Global. doi:10.4018/978-1-5225-5216-1.ch012

Singh, S., Prakash, C., & Uthayakumar, M. (2019). Recent Advancements in Customized Investment Castings Through Additive Manufacturing: Implication of Additive Manufacturing in Investment Casting. In M. Uthayakumar, S. Raj, T. Ko, S. Kumaran, & J. Davim (Eds.), *Handbook of Research on Green Engineering Techniques for Modern Manufacturing* (pp. 24–48). Hershey, PA: IGI Global. doi:10.4018/978-1-5225-5445-5.ch003

Takahashi, Y. (2017). Eco-Friendly On-Site Water Analyses for Ultra-Trace Harmful Ions. In T. Kobayashi (Ed.), *Applied Environmental Materials Science for Sustainability* (pp. 313–326). Hershey, PA: IGI Global. doi:10.4018/978-1-5225-1971-3.ch014

Taokaew, S., Phisalaphong, M., & Newby, B. Z. (2017). Bacterial Cellulose: Biosyntheses, Modifications, and Applications. In T. Kobayashi (Ed.), *Applied Environmental Materials Science for Sustainability* (pp. 255–283). Hershey, PA: IGI Global. doi:10.4018/978-1-5225-1971-3.ch012

Trang, T. T. (2017). Study on the Use of Biomass Polymer Sheets in Water/Alcohol Pervaporation as a Sustainable Source of Alcohol Energy. In T. Kobayashi (Ed.), *Applied Environmental Materials Science for Sustainability* (pp. 234–254). Hershey, PA: IGI Global. doi:10.4018/978-1-5225-1971-3.ch011

Tugsal, U. M., & Taskin, B. (2017). Numerical Methods for the Seismic Performance Assessment of Reinforced Concrete Buildings. In V. Plevris, G. Kremmyda, & Y. Fahjan (Eds.), *Performance-Based Seismic Design of Concrete Structures and Infrastructures* (pp. 275–294). Hershey, PA: IGI Global. doi:10.4018/978-1-5225-2089-4.ch011

Uthayakumar, M., Karnan, B., Slota, A., Zajac, J., & Davim, J. P. (2019). Performance Study of LaPO4-Y2O3 Composite Fabricated by Sol-Gel Process Using Abrasive Waterjet Machining. In M. Uthayakumar, S. Raj, T. Ko, S. Kumaran, & J. Davim (Eds.), *Handbook of Research on Green Engineering Techniques for Modern Manufacturing* (pp. 143–161). Hershey, PA: IGI Global. doi:10.4018/978-1-5225-5445-5.ch009

Uzun, F. V., & Somuncu, M. (2018). Cultural Landscape: An Evaluation From Past to Present. In I. Rosa, J. Lopes, R. Ribeiro, & A. Mendes (Eds.), *Handbook of Research on Methods and Tools for Assessing Cultural Landscape Adaptation* (pp. 1–27). Hershey, PA: IGI Global. doi:10.4018/978-1-5225-4186-8.ch001

Valente, M., & Milani, G. (2017). Seismic Assessment and Retrofitting of an Under-Designed RC Frame Through a Displacement-Based Approach. In V. Plevris, G. Kremmyda, & Y. Fahjan (Eds.), *Performance-Based Seismic Design of Concrete Structures and Infrastructures* (pp. 36–58). Hershey, PA: IGI Global. doi:10.4018/978-1-5225-2089-4.ch002

Vaz de Freitas, I., Marques, J., Rodrigues, C. A., & Sousa, C. (2018). Urban Landscape Quality Management and Monitoring: A Methodological Proposal to Study the Case of Porto, Portugal. In I. Rosa, J. Lopes, R. Ribeiro, & A. Mendes (Eds.), *Handbook of Research on Methods and Tools for Assessing Cultural Landscape Adaptation* (pp. 396–413). Hershey, PA: IGI Global. doi:10.4018/978-1-5225-4186-8.ch016

Waghmode, M. S., Gunjal, A. B., Bhujbal, N. N., Patil, N. N., & Nawani, N. N. (2019). Eco-Friendly Construction. In G. Koç & B. Christiansen (Eds.), *Reusable and Sustainable Building Materials in Modern Architecture* (pp. 80–92). Hershey, PA: IGI Global. doi:10.4018/978-1-5225-6995-4.ch004

Węglarz, A. (2018). The Concept of Expert System Supporting the Increase of Energy Efficiency in Buildings. In E. Rynska, U. Kozminska, K. Zinowiec-Cieplik, J. Rucinska, & B. Szybinska-Matusiak (Eds.), *Design Solutions for nZEB Retrofit Buildings* (pp. 115–140). Hershey, PA: IGI Global. doi:10.4018/978-1-5225-4105-9.ch005

Wesley, D. T., & Puffer, S. M. (2019). The End of Sand: Confronting One of the Greatest Environmental Challenges of the New Millennium. In G. Koç & B. Christiansen (Eds.), *Reusable and Sustainable Building Materials in Modern Architecture* (pp. 1–27). Hershey, PA: IGI Global. doi:10.4018/978-1-5225-6995-4.ch001

Yagoube, I. I., Ibrahim, T., Nor, N. M., & Nallagownden, P. (2019). Performance of PM Linear Generator Under Various Ferromagnetic Materials for Wave Energy Conversion. In M. Uthayakumar, S. Raj, T. Ko, S. Kumaran, & J. Davim (Eds.), *Handbook of Research on Green Engineering Techniques for Modern Manufacturing* (pp. 113–126). Hershey, PA: IGI Global. doi:10.4018/978-1-5225-5445-5.ch007

Yücel, G. F., Işık, B., & Cömert, N. Z. (2018). An Ecological Assessment Analysis: The Kanlidere River in North Cyprus. In I. Rosa, J. Lopes, R. Ribeiro, & A. Mendes (Eds.), *Handbook of Research on Methods and Tools for Assessing Cultural Landscape Adaptation* (pp. 414–433). Hershey, PA: IGI Global. doi:10.4018/978-1-5225-4186-8.ch017

Zielonko-Jung, K., & Poćwierz, M. (2018). The Impact of Forms of Buildings on the Air Exchange in Their Environment: Based on the Example of Urban Development in Warsaw. In E. Rynska, U. Kozminska, K. Zinowiec-Cieplik, J. Rucinska, & B. Szybinska-Matusiak (Eds.), *Design Solutions for nZEB Retrofit Buildings* (pp. 310–330). Hershey, PA: IGI Global. doi:10.4018/978-1-5225-4105-9.ch014

Zindani, D., Faisal, N., & Kumar, K. (2019). Optimization of Process Parameters for Electro-Chemical Machining of EN19: Using Particle Swamp Optimization. In M. Uthayakumar, S. Raj, T. Ko, S. Kumaran, & J. Davim (Eds.), *Handbook of Research on Green Engineering Techniques for Modern Manufacturing* (pp. 127–142). Hershey, PA: IGI Global. doi:10.4018/978-1-5225-5445-5.ch008

Zindani, D., & Kumar, K. (2018). Industrial Applications of Polymer Composite Materials. In K. Kumar & J. Davim (Eds.), *Composites and Advanced Materials for Industrial Applications* (pp. 1–15). Hershey, PA: IGI Global. doi:10.4018/978-1-5225-5216-1.ch001

Zindani, D., & Kumar, K. (2019). Integrated Manufacturing System for Complex Geometries: Towards Zero Waste in Additive Manufacturing. In M. Uthayakumar, S. Raj, T. Ko, S. Kumaran, & J. Davim (Eds.), *Handbook of Research on Green Engineering Techniques for Modern Manufacturing* (pp. 14–23). Hershey, PA: IGI Global. doi:10.4018/978-1-5225-5445-5.ch002

Zindani, D., Maity, S. R., & Bhowmik, S. (2018). A Decision-Making Approach for Material Selection of Polymeric Composite Bumper Beam. In K. Kumar & J. Davim (Eds.), *Composites and Advanced Materials for Industrial Applications* (pp. 112–128). Hershey, PA: IGI Global. doi:10.4018/978-1-5225-5216-1.ch006

Zinowiec-Cieplik, K. (2018). Integration of Vegetation With Architecture Forms. In E. Rynska, U. Kozminska, K. Zinowiec-Cieplik, J. Rucinska, & B. Szybinska-Matusiak (Eds.), *Design Solutions for nZEB Retrofit Buildings* (pp. 275–296). Hershey, PA: IGI Global. doi:10.4018/978-1-5225-4105-9.ch012

About the Authors

Jahangir Mirza is a senior scientist at Research Institute of Hydro-Québec's Research Facility in Montreal, Québec, Canada. He also worked as a Professor and an Adjunct Professor at Faculty of Civil Engineering, Universiti Teknologi Malaysia (UTM), Johor Bharu, Malaysia and in the Civil Engineering and Applied Mechanics Dept. of McGill University, Montreal, Canada, respectively. He is primarily involved in the APPLIED R & D on materials and their application methods not only to repair, maintain and rehabilitate concrete structures, but also to develop new sustainable products using natural, industrial and agriculture wastes to reduce cost, energy and environmental problems for the existing and future construction of concrete structures. He has authored and co-authored more than 200 technical reports and scientific publications, one book and five book chapters. He is a recipient of +25 national and international awards and honours from Canada, South Korea, Malaysia, Pakistan, U.K. and USA which include 6 gold and one silver medals and The Best Paper award from an International conference. He is a member Editorial Board of a Q1 Journal as well as Member Advisory Board of an international conference. Prof. Mirza has conducted several presentations, upon invitation from many countries of the world. He is also a peer reviewer of several international journals as well as M. Sc. and Ph. D. theses. He is a member of Concrete Society of Malaysia and ex-member of American Concrete Institute, Chemical Institute of Canada, Canadian Dam Safety Association and International Concrete Repair Institute.

Mohd Warid Hussin has recently retired as a senior professor at Faculty of Civil Engineering, Universiti Teknologi Malaysia (UTM), Skudai, Johor. He received his PhD degree from University of Sheffield, UK in 1985. Among the positions held in UTM are: Head of Structures & Materials Department, Deputy Dean (Postgraduate & Research), Deputy Dean (Academic), Head of Construction Focus Group and Director of Construction Technology &

Management Centre (CTMC). He is also a member of IEM, BEM, Charter member of International Ferrocement Society (IFS), Concrete Society of Malaysia (MCSM) and ACI Malaysian Chapter. He is currently a director of a consulting company specializing in C & S, project management and forensic engineering. Prof. Dr. Hussin's field of expertise is in concrete technology of which he specializes in fibre-reinforced concrete and blended cement concrete. His involvement in these areas started in 1980 and since then his success in this field was acknowledged worldwide and he is now a renowned expert at both national and international levels. This recognition has paved the way for his appointment as an expert consultant to many organizations including Construction Industry and Development Board (CIDB) of Malaysia, Institut Kerja Raya Malaysia (IKRAM), Jabatan Kerja Raya Malaysia (JKR), International Ferrocement Information Center (IFIC) of Bangkok and the British Council. Prof. Dr. Hussin's key interest of research is on new innovation using alternative materials and modern technology for the development of the construction industry. His achievements in the area of fibre reinforced concrete (FRC), focus in the use of new materials in concrete namely steel, glass, polypropylene, and other natural fibres. Whilst in the field of blended cement concrete his research priority is on the use of cement replacement materials such as pulverized fuel ash (PFA), palm oil fuel ash (POFA), timber industrial ash (TIA) and ground granulated blast furnace slag (GGBFS). These research products on FRC were patented in 2001. His invaluable contributions in this field have significantly improved the quality and performance of concrete with respect to durability, strength and cost. He is also a keen researcher accomplished several programs such as organizing international courses and community development projects in the country. Prof. Dr. Hussin is active in research on new materials for concrete repairs and has acquired two rewarding research grants from the Construction Industry and Development Board (CIDB) of Malaysia. His success in this research has won him a patent for the new innovative products for concrete repairs. The latest and most recent area of his research is in Geopolymer Concrete-a new type of concrete without using any cement for a more sustainable and green concrete construction. The product Cement Free Concrete (Eco - Green Brick EGB) has won gold medals in several exhibitions held internationally including ITEX and SIIF in Korea. Prof. Dr. Hussin has contributed enormously in the field of modern concrete technology. Besides being a dedicated lecturer, as a professor in the University, he had tirelessly work in search of new materials for the growth of the construction industry, which is crucial for a developing country like Malaysia. His perseverance and dedication should entitle him to be known as a Modern Concrete Technologist of this era. He has published more than 300 papers in this field in conference proceedings and journals and has written several books.

Mohamed Abdel Kader Ismail is currently working as a full professor at Dept. of Civil Engineering, Miami College of Henan University, Kaifeng, China since October 2018. He is also an adjunct professor at School of Civil and Mechanical Engineering, Faculty of Engineering and Science, Curtin University, Bentley, Australia. Before joining Miami College, Prof. Ismail was working as an Associate Professor at Civil and Construction Engineering Department, Faculty of Engineering and Science, Curtin University Malaysia, Sarawak, Malaysia. He received his B. Sc. and M. Sc. Degrees from Alexandria University, Egypt in 1991 and 1996, respectively and his PhD from Nanyang Technological University (NTU), Singapore in 2003. Prof. Ismail teaches undergraduate courses in Civil Engineering Materials, Concrete Technology, Engineering Mechanics, Reinforced Concrete Structural Design, Engineering and Environmental Principles and Theories, Construction Management, Concrete Laboratory, Mechanics of Solids, Structural Analysis and Fluid mechanics. For graduate students, he teaches courses in Advanced Concrete Technology and Advanced Structural Analysis. His research work includes Concrete Technology, Smart Materials in Construction, High Performance Concrete, Durability of Concrete, NDT, Arc Thermal Metal Spray Technology, Sustainable Building Materials and Protection Methods of Reinforced Concrete Structures. Prof. Ismail has extensive teaching and research experiences over 25 years in Canada, South Korea, Malaysia, Egypt, Japan and China. He also had broad experience at Engineering Consultancies and Oil & Gas Firms. Prof. Ismail has published more than 120 papers in referred Journals and International Conferences and 5 Books. He served as a reviewer for many International Journals, Editorial Board Member of few journals and Editor-in-chief of Challenge Journal of Concrete Research Letters (CJCRL). He is a member of ASCE and EES and a Professional Engineer of APEGA-Canada.

Index

H

I

L

M

P

Q

R

S

W

X

www.ingramcontent.com/pod-product-compliance
Lightning Source LLC
Chambersburg PA
CBHW081211180926
51613CB00003B/213
* 9 7 8 1 5 2 2 5 8 3 2 5 7 *